CELL BIOLOGY RESEARCH PROGRESS

CHEMOTAXIS: TYPES, CLINICAL SIGNIFICANCE, AND MATHEMATICAL MODELS

CELL BIOLOGY RESEARCH PROGRESS

Additional books in this series can be found on Nova's website under the Series tab.

Additional E-books in this series can be found on Nova's website under the E-books tab.

CELL BIOLOGY RESEARCH PROGRESS

CHEMOTAXIS: TYPES, CLINICAL SIGNIFICANCE, AND MATHEMATICAL MODELS

TIMOTHY C. WILLIAMS
EDITOR

Nova Science Publishers, Inc.
New York

For permission to use material from this book please contact us:
Telephone 631-231-7269; Fax 631-231-8175
Web Site: http://www.novapublishers.com

Library of Congress Cataloging-in-Publication Data

Chemotaxis : types, clinical significance, and mathematical models / Timothy C. Williams.
p. cm.
Includes index.
ISBN 978-1-61728-495-3 (hardcover)
1. Chemotaxis. I. Williams, Timothy C.
QR187.4.C54 2009
578.01'2--dc22
2010025430

Published by Nova Science Publishers, Inc. † New York

Contents

Preface vii

Chapter 1 The Cell Migration Signalosome 1
Gabriele Eden, Marco Archinti, Federico Furlan, Paul Fitzpatrick, Ronan Murphy and Bernard Degryse

Chapter 2 Plant Growth-Promoting Bacteria: The Role of Chemotaxis in the Association Azospirillum Brasilense-Plant 53
Raúl O. Pedraza, María I. Mentel, Alicia L. Ragout, Ma. Luisa Xiqui, Dulce Ma. Segundo and B. E. Baca

Chapter 3 CD46, Chemotaxis and MS– Are These Linked? 85
Siobhan Ni Choileain, Jillian Stephen, Belinda Weller and Anne L. Astier

Chapter 4 Regulation of Chemotaxis by Heterotrimeric G Proteins 109
Maggie M. K. Lee and Yung H. Wong

Chapter 5 Chemotaxis in the Model Organism *Chlamydomonas Reinhardtii* 135
E.G. Govorunova, O.O. Voytsekh, A.P. Filonova, M.A. Kutuzov, M. Mittag and O.A. Sineshchekov

Chapter 6 Chemotaxis-Based Assay for the Biological Action of Silver Nanoparticles **157**
E. M.Egorova, S. I. Beylina, N. B.Matveeva and L. S. Sosenkova

Chapter 7 Chlamydomonas as the Unicellular Model for Chemotaxis and Cellular Differentiation **189**
E.V. Ermilova

Index **211**

Preface

Chemotaxis is the phenomenon in which bodily cells, bacteria, and other single-cell or multicellular organisms direct their movements according to certain chemicals in their environment. This is important for bacteria to find food by swimming towards the highest concentration of food molecules, or to flee from poisons. In multicellular organisms, chemotaxis is critical to early and subsequent phases of development, as well as in normal function. This new book discusses reserch in the study of chemotaxis including the cell migration signalosome, the role of chemotaxis in the association of the azospirillum brasilense plant, the role of CD46 in the control of chemotaxis of activated T cells in MS pathogenesis and the regulation of chemotaxis by heterotrimeric G proteins.

Chapter 1- Cell migration is an essential process involved in the development of the organism during embryogenesis and then throughout life in physiological and pathological events including inflammation, wound healing, vascular diseases and cancers. The urokinase receptor (uPAR), a glycosyl-phosphatidyl-inositol (GPI) anchored-protein, regulates a variety of important cellular processes such as pericellular proteolysis and cell migration. In fact, uPAR is involved in different kinds of cell migration including haptotaxis, diapedesis and chemotaxis. uPAR is an important motogenic receptor because it can directly regulate and coordinate not only the generation of the signal that actually starts the migration of the cell but also the degradation of the proteins of the extracellular matrix that opens the path to the migratory cells. In addition, uPAR has other unique properties resulting from its relationship with a wide array of membrane receptors such as seven-transmembrane domain receptors, tyrosine kinase receptors and integrins. By initiating the formation of large signalling complexes, uPAR is the foundation stone of the cell migration signalosome primarily constituted by uPAR and the integrins. These latter compose a large family of membrane receptors that provides connection to the cell cytoskeleton, cell survival and adhesion. Thereby, the combination of uPAR and the integrins brings new essential properties to the migratory cell. Furthermore, the exceptionally wide array of ligands and membrane-bound partners of uPAR and the integrins consents to add new constituents, finely adjust the composition, and modulate the functions of the cell migration signalosome in order to meet the various cellular requirements of the migratory cell throughout its journey. These properties are perfectly suited for the spatial and temporal regulation of cell migration.

This chapter will discuss the molecular mechanisms of cell migration with a particular emphasize on the role of uPAR and uPAR-integrin interactions that lead to the formation of the cell migration signalosome.

Chapter 2- The genus *Azospirillum* belongs to the plant growth-promoting bacteria group, capable of positively influencing the growth and yield of numerous plant species, many of them with agronomic and ecological importance. Plant growth promotion is largely determined by the efficient colonization of the rhizosphere (soil influenced by roots and microorganisms). Root exudates constitute the most significant source of nutrients in the rhizosphere and seem to participate in the early colonization by inducing the chemotactic response of bacteria. Therefore, chemotaxis is considered an essential mechanism for the successful root colonization by *Azospirillum.*

In this chapter the authors present a background and new insights on *Azospirillum* chemotaxis, concerning the genetic aspects and its use for addressing biotechnological applications. First, the authors demonstrate that the genetic complementation of a mutant strain, impaired in surface motility, led to the identification of the gene *chsA* (<u>ch</u>emotactic <u>s</u>ignaling protein). The deduced translation product, ChsA protein, contained a PAS sensory domain and EAL active site domain. The latter has phosphodiesterase activity (PDE-A) for the hydrolysis of c-di-GMP [cyclic-bis (3' –5') dimeric GMP], a compound known to function as a second messenger in different cellular processes, including motility, biofilm formation and cellular differentiation.

After cloning *chs*A, ChsA protein was expressed and purified by affinity chromatography. ChsA activity in presence of *bis-p*-nitro phenyl-phosphate was 0.59-ÂμM min^{-1} mg^{-1} protein, demonstrating that it displayed phosphodiesterase activity. This suggests that ChsA is a component of the signaling pathway controlling chemotaxis in *Azospirillum.* Then, the authors propose that the redox state of the cell is sensed through the PAS domain and directly coupled to the transmitter EAL module, showing PDE-A activity.

Second, the chemotaxis of different strains of A. brasilense toward strawberry root exudates was investigated. The agar-plate assay was used, including two concentrations of exudates from three commercial varieties of strawberry, collected at different time intervals. To quantify the chemotactic response, the capillary method was used. In all cases, a positive chemotactic reaction was found, revealing higher responses in endophytic than in rhizospheric strains, being this strain-specific. Furthermore, the variation of the chemotactic response observed depended on the concentration and time to collect the exudates, as well as the total sugar content. Considering that A. brasilense possessses biotechnological application, addressing to a sustainable agriculture, determining the genes and mechanisms involved in chemotaxis response, as well as the level of activity of strains to root exudates may represent an initial step in selecting them for use as inoculants in different crops.

Chapter 3- Multiple Sclerosis (MS) is an autoimmune disease characterized by chronic inflammation of the brain. One of the main occurrences is the breach of the blood-brain-barrier, resulting in the entrance of inflammatory cells, which perpetuate the inflammation occurring in the brain. One of the mechanisms controlling cell migration is mediated by the release of small soluble molecules, called chemokines and by the expression of their corresponding receptors, the chemokine receptors. Hence, the conjoint expression of chemokine receptors and production of their relevant chemokines will direct the migration of cells towards the site of inflammation. Among the cells involved in the pathogenesis of MS, T cells have been shown to play an important role. Indeed, T cell activation is crucial for the immune homeostasis, notably through the balance of effectors T cells (Teff) and regulatory T cells (Tregs). In MS, defective Treg functions have been observed, which might partly explain the increased inflammation seen in MS. Among Tregs, Tr1 cells are characterized by the secretion of large amount of IL-10, an anti-inflammatory cytokine. The molecule CD46 is a regulator of complement activity. However, its activation also promotes T cell activation and differentiation toward Tr1 cells. This pathway is altered in MS, as the amount of IL-10 produced by CD46-activated T cells is largely reduced. This chapter will discuss preliminary evidence suggestive of a role of CD46 in the control of chemotaxis of activated T cells, which might play a role in MS pathogenesis.

Chapter 4- With the ability to mediate chemotaxis of inflammatory cells, chemoattractants have been shown to contribute to the development of inflammatory diseases, such as atherosclerosis and angiogenesis. Many chemoattractant receptors belong to the family of seven-transmembrane-domain G protein-coupled receptors (GPCRs) and elicit their effects through heterotrimeric (αβγ) guanine nucleotide-binding proteins (G proteins). The three subunits form two functional compartments – the Gα subunit and the stable Gβγ complex. Both the dissociated GTP-bound Gα subunit and Gβγ complex can exert biological effects. G proteins are classified into four major subfamilies, G_s, G_i, G_q and G_{12}, according to the amino acid sequence homology and functional specialization of the Gα subunit. Members in all G protein subfamilies are known to interact with chemoattractant receptors individually or simultaneously, which trigger the activation of multiple signaling molecules, leading to actin reorganization and subsequent cell mobilization. In this chapter, the authors will discuss the promiscuity of chemoattractant receptors in G protein coupling and examine their underlying molecular mechanisms in directed cell migration.

Chapter 5- The unicellular biflagellate alga *Chlamydomonas reinhardtii* is widely used as a simple model to study fundamental cellular processes, including chemotaxis. *C. reinhardtii* is attracted towards ammonium ions (NH_4^+) and peptide mixtures, such as a pancreatic digest of casein (tryptone). The sensitivity to NH_4^+ is transiently induced in vegetative cells by nitrogen deprivation, whereas the sensitivity to tryptone requires formation of mature gametes. The clock-controlled RNA-binding protein CHLAMY1 might be involved in regulation of the diurnal rhythm of chemotaxis to NH_4^+. The authors measured inhibition of rhodopsin-mediated photoreceptor currents by the chemoattractant tryptone as an indirect assay of early stages of chemosensory transduction in *C. reinhardtii* gametes. The results showed that the magnitude of the response to tryptone depended on the concentration of monovalent metal cations in the medium with the selectivity sequence K^+>Rb^+>Cs^+>Na^+>Li^+. It was inhibited by extracellular Ba^{2+} and Ca^{2+} in millimolar concentrations, and by dibutyryl-cAMP. These observations suggest the involvement of K^+ channels modulated by cyclic nucleotides in *C. reinhardtii* chemotaxis. In order to identify active ingredients, the authors subjected tryptone to fractionation by gel filtration and demonstrated that only fractions that contain individual amino acids and dipeptides were functionally active. However, none of the 18 tested individual amino acids fully mimicked the effect of tryptone. The authors hypothesize that a specific combination of amino acids and/or dipeptides acts as a chemoattractant for *C. reinhardtii* gametes.

Chapter 6- The authors propose a new way for testing the biological action of silver nanoparticles. The nanoparticles (with sizes ranging from 3 to 16 nm) were obtained by the original method of biochemical synthesis in reverse micelles stabilized by anionic surfactant bis-(2-ethylhexyl) sodium sulfosuccinate (AOT). From micellar solution the nanoparticles were transferred into the water phase; water solutions of the nanoparticles were used for testing their biological activity. Our assay is based on negative chemotaxis, a motile reaction of cells to an unfavorable chemical environment. Plasmodium of the slime mold *Physarum polycephalum* used as an object is a multinuclear amoeboid cell with unlimited growth and auto-oscillatory mode of locomotion. Biocidal and repellent effects were compared for silver nanoparticles, Ag^+ ions, and AOT; the latter two reagents were introduced in the concentrations equal to those present in the nanoparticles' solution. All substances were tested in water solution and in the agar gel. The authors have revealed that in characteristics common for repellents, such as increase of the period of contractile auto-oscillations, decrease of the area of spreading on substrate, and substrate preference in spatial tests, silver nanoparticles proved to be substantially more effective than Ag^+ ions, AOT and the sum Ag^+ + AOT. The lethal concentration of the nanoparticles for macroplasmodium in water solution was found to be about 10 μg/ml, the concentrations effective for chemotaxis were 30 times lower. The chemotactic tests allow the quantitative estimation of the biological reaction and monitoring of its dynamics; in resolution, they are superior to the tests based on the lethal action of biocidal agents. It is shown also that the spatial chemotactic tests are sensitive enough to distinguish between the effectiveness of different nanoparticle preparations. In particular, the authors found that, at the equal silver concentration, the repellent activity is higher for the 5 nm than for the 9 nm particles.

The results obtained allow to conclude that the chemotaxis-based assay could be helpful for finding a proper balance between the efficiency of silver nanoparticles as antimicrobial drug and the risk of tissue damage during medical treatment, and, hence, for elaborating an optimal protocol of their clinical application.

Chapter 7- *Chlamydomonas* has long been one of the most successful unicellular organism for genetic and biochemical studies of the photosynthesis, organelle genomes and flagellar assembly. The availability of the new molecular genetic techniques is increasing interest in *Chlamydomonas* as a model system for research in areas like swimming behavior where it previously has not been widely exploited. The swimming behavior of *Chlamydomonas reinhardtii* is influenced by several different external stimuli including chemical attractants. Chemotaxis of the green alga is altered during gametic differentiation. Gametogenesis results in the conversion of chemotactically active vegetative cells into chemotactically inactive gametes. This experimental system offers the opportunity to study cellular behavior and differentiation at the molecular level with use of a wide range of molecular genetic approaches, including gene tagging by insertional mutagenesis, quantitative PCR and RNA interference. In this chapter I discuss recent progress in the field of chemotaxis in *Chlamydomonas*. Emphasis is placed on the signal pathways by which the two environmental cues – ammonium and light control chemotaxis and gametic differentiation.

In: Chemotaxis: Types, Clinical Significance... ISBN: 978-1-61728-495-3
Editor: Timothy C. Williams, pp. 1-52

Chapter 1

The Cell Migration Signalosome

Gabriele Eden[1], Marco Archinti[2], Federico Furlan[3], Paul Fitzpatrick[4], Ronan Murphy[4] and Bernard Degryse*

[1]Division of Nephrology, Hannover Medical School, Hannover, Germany

[2]Institute for Research in Biomedicine, Parc Cientific de Barcelona, Baldiri Reixac 10, 08028 Barcelona, Spain

[3]BoNetwork Programme, San Raffaele Scientific Institute. Milan, Italy

[4]School of Health and Human Performance, Faculty of Science and Health, Dublin City University, Glasnevin, Dublin 9, Ireland

Abstract

Cell migration is an essential process involved in the development of the organism during embryogenesis and then throughout life in physiological and pathological events including inflammation, wound healing, vascular diseases and cancers. The urokinase receptor (uPAR), a glycosyl-phosphatidyl-inositol (GPI) anchored-protein, regulates a variety of important cellular processes such as pericellular proteolysis and cell migration. In fact, uPAR is involved in different kinds of cell migration including haptotaxis, diapedesis and chemotaxis. uPAR is an important motogenic receptor because it can directly regulate and coordinate not only the generation of the signal that actually starts the migration of the cell but also the degradation of the proteins of the extracellular matrix

* Correspondig author: E-mail: bdegryse@yahoo.com

that opens the path to the migratory cells. In addition, uPAR has other unique properties resulting from its relationship with a wide array of membrane receptors such as seven-transmembrane domain receptors, tyrosine kinase receptors and integrins. By initiating the formation of large signalling complexes, uPAR is the foundation stone of the cell migration signalosome primarily constituted by uPAR and the integrins. These latter compose a large family of membrane receptors that provides connection to the cell cytoskeleton, cell survival and adhesion. Thereby, the combination of uPAR and the integrins brings new essential properties to the migratory cell. Furthermore, the exceptionally wide array of ligands and membrane-bound partners of uPAR and the integrins consents to add new constituents, finely adjust the composition, and modulate the functions of the cell migration signalosome in order to meet the various cellular requirements of the migratory cell throughout its journey. These properties are perfectly suited for the spatial and temporal regulation of cell migration.

This chapter will discuss the molecular mechanisms of cell migration with a particular emphasize on the role of uPAR and uPAR-integrin interactions that lead to the formation of the cell migration signalosome.

Introduction

Cell migration is one basic but important process, in fact cell migration can be viewed as the grand architect of the organism [for reviews see Horwitz and Webb, 2003; Ridley et al., 2003; Bagorda et al., 2006; Gerthoffer, 2008; Bretscher, 2008]. During embryogenesis, cell migration accompanies differentiation that sees the three original layers of cells, the ectoderm, mesoderm and endoderm generating all organs of the organism and shaping the morphology of each individual. After birth, cell migration wears several hats. As security officer, cell migration leads the defense. Its duties are to serve and protect being involved in the patrols and when needed in the recruitment of the immune cells. In addition, cell migration is also the chief mechanic supervising body repair. Wound healing is a really complex process requiring the cleaning of cellular debris, removal of possible infectious agents, migration of the right type of cells that will close the wound, and the reorganization/reformation of the blood vessel network. Unfortunately, the dark side of the process is painted in deep black. Cell migration is involved in many pathological states comprising vascular diseases and cancers that are major causes of death in western societies. Inflammation marked by redness, heat, swelling and pain is another well known example. These brief lines plainly justify the effort of research on cell migration that may bring new understanding of the rise and fall of life [Horwitz and Webb, 2003; Ridley et al., 2003; Bagorda et al., 2006; Gerthoffer, 2008; Bretscher, 2008].

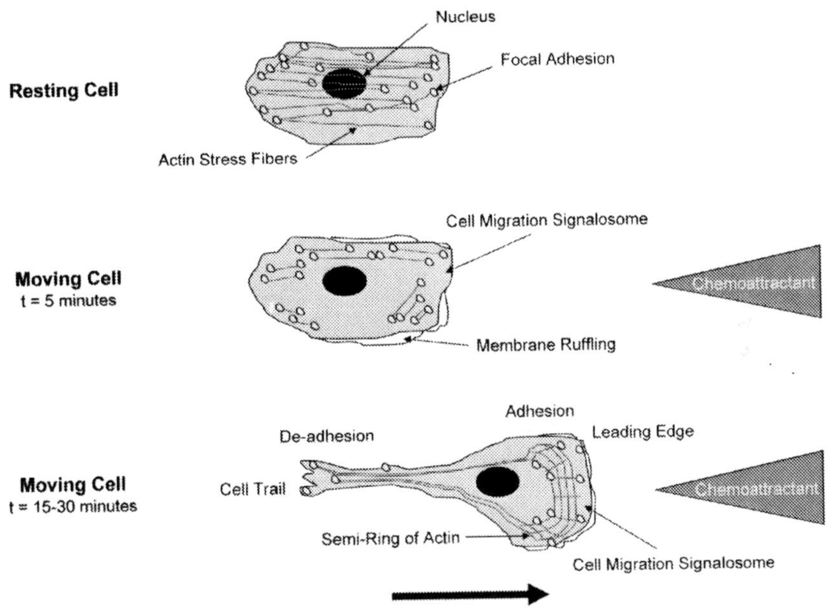

Figure 1. Model of mesenchymal cell migration

Nevertheless, other modes of cell migration have been described. Neuron cells have been showed migrating in the developing cerebral cortex as branching cells exhibiting a long branched extension that senses the ECM and governs the direction of cell migration [Nadarajah et al., 2003]. A function very similar to that of the cellular protrusions acting as sticky fingers observed in other cells [Galbraith et al., 2007]. These cellular "stinky fingers" probe the ECM and control the direction of cell migration [Galbraith et al., 2007]. Other neurons exhibit a similar motile morphology but with the neurite extension that is rather used for nucleokinesis [Nadarajah et al., 2003]. Nucleokinesis is an interesting mode of cell migration that consists of three main steps: the extension of the leading neurite ahead of the cell body, the relocalization of the centrosome or main microtubule-organizing centre (MTOC) in the leading neurite, and finally the nuclear movement that follows the MTOC. In addition, while the MTOC appears to glide very smoothly forward, the nucleus follows by successive jerky jumps [for review see Burke and Roux, 2009].

Table 1. Short lexicon of the types of cell migration

Type of cell migration	Definition
Chemotaxis	Cell migration in a specific direction controlled by a gradient of a diffusible agent (chemoattractant)
Haptotaxis	Cell migration directed by a gradient of substrate-bound attractant
Durotaxis or Mechanotaxis	Cell migration directed by a gradient of stiffness of the substrate or by the biomechanical forces
Chemokinesis	Migration of cells induced by an attractant but in random direction because of the absence of gradient
Nucleokinesis	Typical migration of bipolar neuronal cells which is characterized of the migration of the nucleus behind a long leading cellular extension
Diapedesis	extravasation of leukocytes across the endothelial wall of a blood vessel
Extravasation	Migration of leukocytes or tumour cells out of the circulation into a tissue
Intravasation	Migration of cells from the tissue into the circulation
Amoeboid	Protease-independent migration of cells moving by rapid cycles of morphological expansion and contraction
Fibroblast-like or Mesenchymal	Protease-dependent migration of cells moving by assembling/disassembling adhesive contacts with the substrate and transmitting force from these adhesion sites to the substrate
Chain Migration or Cellular Streaming	Multi cell migration mode in which cells migrate as a single file in a head-to-tail fashion behind a cellular leader
Cluster Migration	Multi cell migration mode in which aggregated cells move behind a leading front composed of cells exhibiting numerous cellular protrusions

The plasminogen activation system and in particular the urokinase receptor (uPAR) is known to be involved in cell migration [for reviews see: Blasi and Carmeliet, 2002; Degryse, 2003, 2008; Ragno, 2006; Archinti et al., 2010]. An impressive body of evidence shows that uPAR is a motogenic receptor mediating migratory signals to the cell thereby starting cell motility. Furthermore, uPAR provides unique advantages to the migrating cell. By controlling pericellular proteolysis, uPAR opens the path and severs cell-cell and cell-matrix contacts that otherwise should impede cell migration. uPAR is also an adhesion receptor that keeps the cells attached to the substrate avoiding anoikis triggered by the loss of cell-extracellular matrix (ECM) contacts. In addition, by interacting with the large family of adhesion receptors, the integrins, uPAR promotes the formation of focal complexes that mature into focal adhesions [Blasi and Carmeliet, 2002; Degryse, 2003, 2008; Ragno, 2006; Archinti et al., 2010].

The relationship with the integrins is of exceptional importance because the combination of the uPAR system with the integrins brings ultimate advantages to the migratory cells: direct connection with the cell cytoskeleton, mechanotransduction, cell survival and resistance to genotoxic injury [Degryse, 2008]. Furthermore, due to the numerous ligands and lateral partners of uPAR and integrins, the formation of uPAR-integrin complex lays the foundation stone of a larger and modular signalling complex i.e. the cell migration signalosome. The flexibility in the molecular composition of this complex insures meeting the various needs of the migrating cell throughout its whole journey. Thus, the cell migration signalosome constitutes a remarkable adaptable signalling complex i.e. a perfect chameleon signalosome which is the particular topic of this chapter [Degryse, 2008].

Cell Migration

A Touch of History

The first use of the word cell is commonly attributed to the work of Robert Hooke published in micrographia (1665) [Hooke, 1665]. However, the definition of the cell given by Robert Hooke was different from the modern definition of the cell. The concept that cells are the basic component of plants and animals is due to Matthias Jakob Schleiden (1838) and Theodor Schwann (1839) [Schleiden, 1838; Schwann, 1839; for a recent review see Mazzarello, 1999]. Thus, the idea that cells are capable of motility certainly arose very soon thereafter. To our knowledge, in the XIXth century research was already performed on cell migration and reports were either orally presented and/or published [Waller, 1846; Caton, 1870; Felts 1870; Hayem, 1870]. During this period, scientists seem to have been aware of the migration of immune cells to the site of inflammation, and questioning the mechanism of cell migration. 140 years later, incredible progresses have been made towards the elucidation of this mechanism, but despite all our research and technological advances the mechanism of cell migration is still under investigation. It is some kind of ironic that the determination of the mechanism of diapedesis which was the goal of the study of Dr. Caton (1870) is still elusive. Yet, we do not exactly know whether an immune cell goes in between the cells (paracellular) or through one cell of the endothelial wall (transcellular) [Sage and Carman, 2009].

A Definition of cell Migration

The definition of cell migration seems quite obvious, cell migration is the movement of cell(s) from one to another location, but in reality this apparent simplicity conceals an extreme complexity. For instance, migration of neuron cell may refer to the migration of the whole cell during embryogenesis but may also define the extension of the axon without the movement of the cell body. Moreover, this term embraces diverse types of migration as cells can migrate isolated or in group. Unfortunately, just to step up in complexity various sub-types have been described for individual and collective cell migration (Table 1).

An individual cell can move assuming either the fibroblast-like or amoeboid migration which are the best characterized modes of single cell migration [for reviews see Friedl et al., 2001; Horwitz and Webb, 2003; Ridley et al., 2003; Bagorda et al., 2006; Smirnova and Segall, 2007; Gerthoffer, 2008; Bretscher, 2008; Lämmermann and Sixt, 2009; Friedl and Wolf, 2010]. The classical fibroblast-like or mesenchymal migration needs the attachment/detachment of the cell via adhesive sites to the substrate (Figure 1). The cell assumes an elongated spindle-shaped morphology often called hand-mirror shape with a large head at the cell front and a thin trail (the uropod) at the cell rear (Figure 1). The mesenchymal type of cell migration is also dependent on pericellular proteolysis and on the transmission of the biomechanical forces via the actin cytoskeleton from the adhesion sites to the substrate. The amoeboid migration neither requires strong cell adhesion to the substrate nor an active system of pericellular proteolysis. The cell moves rapidly by cycles of morphological expansion and contraction. However, the amoeboid migration ranges from the simple blebbing mode to the more complex gliding mode which is dependent on actin cytoskeleton elasticity. The main differences between these mesenchymal and amoeboid cell migration reside in the speed of cell locomotion, low in the mesenchymal versus high in the amoeboid migration. This difference in velocity conversely correlates with the adhesion strength to the substrate strong in the mesenchymal versus weak in the amoeboid motility [Friedl and Wolf, 2010]. In addition, the integrins are organized in focal complexes and adhesions in the former while the integrins are distributed in diffuse pattern in the latter sub-type of cell migration [Wolf et al., 2003]. The dependence on the protease activity is also distinct. The mesenchymal migration has the need for pericellular proteolysis suggesting that in this kind of motility the cell opens its path across the tissue and the ECM, while the amoeboid does not require an active pericellular proteolysis leading to think that the cell squeeze through tissular and ECM gaps [Wolf et al., 2003].

Kinetic of events occurring in cell migration in two-dimensional culture after the addition of a chemoattractant. Top panel, the resting cell exhibits numerous stress fibers (red filaments) and evenly distributed adhesive sites (yellow focal adhesions or adhesion plaques in the smooth muscle cells). The Cell is not polarized. Middle panel, immediately after the addition of the chemoattractant the migratory signal is received by the signalling receptors, and the cell starts responding. As shown here, after 5 minutes the number of stress fibers is considerably reduced, focal adhesions are redistributed at the periphery of the cell, and membrane ruffling is visible. The cell is creating a leading edge by the accumulation of actin on one side of the cell, and the cell migration signalosome is assembled on this side. Bottom panel, after 15-30 minutes the cell is now polarized displaying a typical motile morphology with the leading edge and the thin cell trail (this morphology is often named hand-mirror shape). The actin cytoskeleton is completely reorganized with a semi-ring of actin at the leading edge. The distribution of focal adhesions in double rows on each side of the semi-ring shows that there is an active process of adhesion at the leading edge of the cell. The generation of new adhesion contacts with the substrate promotes the cellular movement forward and results in the formation of the motile cell morphology. The release of the adhesion attachments at the rear and the contraction of the cell rear contribute to the cell locomotion forward. The cell migration signalosome is located at the leading edge and involved in the regulation of cell motility. Thus, several complicated biological processes occurs simultaneously in the motile cell including reception and transmission of the migratory signal, formation of cellular protrusions (ruffles depicted here), activation of pericellular proteolysis, adhesion at the cell front and de-adhesion at the cell rear, reorganization of the cytoskeleton (only actin is shown here but other components such as microtubules are reorganized) which also serves to the transmission of the biomechanical forces, and generation of energy required for the motility. The large black arrow below the cell indicates the direction of cell migration. Actual pictures of migrating cells (mainly rat smooth muscle cells) can be viewed in previous reports [Degryse et al., 1999, 2001a,b,c, 2004, 2005; Kamikubo et al., 2009; Archinti et al., 2010].

Embryogenesis and in particular the phase of differentiation, is certainly the period of the life of an organism that sees the most important and intense waves of collective cell migration that will ultimately result in the formation of the organs [Yin et al., 2009; Weijer, 2009]. However, multi cell migration also takes place after that period. Collective cell migration is believed to be involved in wound healing but also in tissue invasion by tumoral cells [for reviews see Santoro and Gaudino, 2005; Rørth, 2009; Ilina and Freidl, 2009]. Multi cell migration occurs through two majors modes. On the one hand, the chain migration or cellular streaming in which the cells migrate as a single file in a head-to-tail fashion behind a cellular leader [Davis and Trinkaus, 1981; Teddy and Kulesa, 2004; Rørth, 2009; Friedl and Wolf, 2010]. On the other hand, the cluster migration that sees the locomotion of aggregated cells behind a large leading front composed of several cells exhibiting a motile phenotype similar to the leading edge of a single motile cell with cellular protrusions such as lamellipodia. In the motile cluster, cells at the rear look more like the trail of an individual motile cell, while the cells within the cluster maintain the phenotype of non-motile cells [Horwitz and Webb, 2003; Weijer, 2009; Friedl and Wolf, 2010].

Interestingly, switches from one mode to another mode of cell migration are possible, and these various modes may represent variations of a common theme [Friedl and Wolf, 2003, 2010; Horwitz and Webb, 2003; Wolf et al., 2003]. Indeed, whatever the mode of migration a cell is using, the physical parameters that have to be fulfilled for the cell locomotion are always the same. The diverse mode of cell migration may simply represent the best compromise fitting these physical parameters in a particular situation i.e. a specific tissue environment. In addition, the different motile phenotypes may also be useful for the modulation and the spatial location of particular intracellular signalling pathways [Meyers et al., 2006]. Conversely, intracellular signalling may also influence the type of cell migration used by the cell. Intracellular signalling is indeed crucial for cell migration [for reviews see Ridley et al., 2003; Van Haastert and Veltman, 2007]. For our part, we showed that wild-type fibroblasts migrated as single cell, whereas *c-Src*$^{-/-}$ fibroblasts moved using the cluster type of cell migration [Degryse et al., 1999].

The transition between one mode and the other seems to be governed by both extracellular and intracellular factors [Wolf et al., 2003; Friedl and Wolf, 2003, 2010]. Mathematical models can be used to predict the migration pattern influenced by the adhesion substrate (ECM protein or other cells) and the intra- or inter-cellular communications [Aubert et al., 2008]. Recently, the tuning model of cell migration has been proposed to justify the type of migration that the cell adopts in the presence of specific external and internal determinants [Friedl and Wolf, 2010].

The Cell Migration Cycle

Cell migration can be conveniently represented as a cycle, which helps to understand that migration is a finely coordinated process. However, if the cell migration cycle is an excellent model of the process of cell locomotion, one should keep in mind that some steps of cell migration cycle can happen simultaneously (but at different cellular locations) rather than successively (Fig.1). Also, the model seems to apply better to slow moving cells such as fibroblasts rather than to faster cells such as leukocytes [Horwitz and Webb, 2003; Ridley et al., 2003]. Briefly, the cell migration cycle start with the formation of cellular protrusions upon the reception of a migratory signal, the generation of new adhesion contacts with the substrate that promotes the cellular movement forward and results in an elongated cell morphology, the release of the adhesion attachments at the rear and the contraction of the cell rear contribute to the cell locomotion forward (Fig.1). Then, the cycle revolves once again by the creation of new adhesion sites. Below, we discuss the various steps of the cell locomotion.

Signalling Starts Cell Migration

Cell migration is usually started by a migratory signal. There is an extreme variety among the numerous motogenes so far described which for example comprise nucleotides such as ATP and UTP; formylated peptides such as fMLP; chemokines and CLF chemokines; growth factors; ECM proteins such as vitronectin (VN) and fibronectin (FN); proteases such as thrombin and urokinase (uPA), force or biomechanical stimuli, etc… As these few examples obviously show, there is no shared structure among the diverse motogenic factors. However often two major groups of receptors are activated by the motogenes: the G protein coupled receptors (GPCRs) or seven transmembrane domain receptors (7TMRs), and the growth factor receptors or receptor tyrosine kinases (RTKs). If signalling is the common step initiating cell locomotion, the next events are diametrically opposed. Adherent cells start to de-adhere severing or deactivating cell-cell and cell-substrate contacts such as desmosomes and hemidesmosomes, whereas non-adherent cells adhere to the substrate, an essential process controlled by the integrin family of receptors. The rolling phase of the leukocytes is a well-known example [Schober and Weber, 2005; Fernandez-Borja et al., 2010].

Then, in both adherent and non-adherent cells the course of events tends to be again convergent with an impressive reorganization of the cell cytoskeleton regulated by small GTP binding proteins including Rho and Cdc42, which also embraces the generation of cellular protrusions [Charest and Firtel, 2007; Szczepanowska, 2009]. These protrusions can be as different as blebs, membrane ruffling, lamellipodia, filopodia, pseudopodia, micro-spikes, podosome, invadopodia, neurites, branched processes [Jiang, 1995; Nadarajah et al., 2003; McNiven et al., 2004; Le Clainche and Carlier, 2008; Olson and Sahai, 2009; Albiges-Rizo et al., 2009]. However, all these protrusions are linked to cell movement, and most seem to be particularly involved in the creation of new adhesives sites. Behind these protrusions that mark the very leading edge of the motile cell, the actin cytoskeleton is reorganized in order to transmit the forces needed for the physical locomotion of the cell. The smooth muscle cells (SMC) exhibit a spectacular semi-ring of actin in between a double row of focal adhesions [Degryse et al., 1999]. This semi-ring of actin can also be observed in other cells [Olson and Sahai, 2009]. In addition, the SMC can also show a remarkable network of actin around the cell nucleus that may serve to pull the nucleus during cell locomotion [Degryse et al., 1999]. However, regulation of the shape of the nucleus by mechanotransduction is also involved in gene regulation [Wang et al., 2009].

Biomechanical Force

The correct transmission of the tractional force is essential for cell migration. Moreover, force can also initiate cell locomotion, a process called mechanotaxis [Li et al., 2005]. The network of actin which is always in polymerization/depolymerization is a convenient plastic network for force transmission. External forces get in touch with the actin cytoskeleton through mechanoreceptors. External force promotes the formation of focal complexes that mature into focal adhesions which are clusters of integrins and uPAR. Interestingly, the integrins and uPAR, which are the foundation stones of the cell migration signalosome that we discuss below, are biomechanical sensors and mediators of biomechanical forces, translating them into biochemical signals [Wang et al., 1995; Silver and Siperko, 2003; Katsumi et al., 2004; Katoh et al., 2008; Chen, 2008; Gieni and Hendzel, 2008].

The Polarization of The Motile cell

The formation of the leading edge of the motile cell reveals that an important step has been made i.e. the polarization of the cell affecting the whole cell machinery. This requisite step sees the formation of the elongated motile cell morphology with a large part that forms the leading front of the cell with its cellular protrusions at the vanguard, and a thin elongated trail. In early studies on lymphocytes the name of hand-mirror shape was given to that specific morphology [Gormley and Ross, 1972; Schreiner and Unanue, 1975; Norberg et al., 1978; Thomas et al., 1982; Liso et al., 1983; Schmitt-Gräff and Fischer, 1983]. The whole cell cytoskeleton not only the actin filaments but also the microtubules is reorganized for the cell movement and reflects this polarization. In addition, a wide array of signalling molecules are activated and/or deactivated, and are also redistributed within the motile cell. The receptors and the various other effectors such as kinases, adaptors, and small GTP binding proteins that are involved in the sensing and mediation of the migratory signal accumulate at the leading part of the cell [Lauffenburger and Horwitz, 1996; Ridley et al., 2003; Degryse et al., 1999, 2004; Vicente-Manzanares et al., 2005]. Signalling mechanisms that will prevent the desentization of the cell, thereby avoiding that the cell lost its track will be also activated. Ca^{2+} oscillations have been suggested to avoid desentization to the attractant [Liu et al., 2009]. In the opposite compartment, the cell rear, other factors such as myosin II that are rather responsible of the retractation of the cell rear relocalize in the uropod [Lauffenburger and Horwitz, 1996; Ridley et al., 2003; Vicente-Manzanares et al., 2005]. Endocytosis of some components of the focal adhesions including the integrins participates in the disassembly of the adhesive contacts [Ezratty et al., 2005]. Integrin deactivation also takes place in the cell rear. If the integrins are not timely deactivated the locomotion is doomed to fail. The cell will eventually elongate in an effort to move forward but will not actually go further, and soon the cell will be abruptly pulled back by the still adhesive integrins [for a spectacular illustration see the supplementary movies of Zijlstra et al., 2008]. The tetraspanin CD151 was shown to be in charge of integrin deactivation [Zijlstra et al., 2008]. Interestingly, talin which has been very recently shown to be sufficient for integrin activation [Ye et al., 2010], is proteolytically degraded reinforcing the irreversible status of the dismantlement of the focal adhesions [Franco et al., 2004].

The cell is now ready for the move.

The Cell Migration Signalosome

This part of this chapter is particularly dedicated to the description of the cell migration signalosome, the macromolecular complex that leads and coordinates cell motility.

The Walk on the Integrins

Just like anyone walking on earth, cells generate adhesion at the front and de-adhesion at rear. However, the extraordinary shoes of the cell are made of integrins. Moreover, integrins constitutes part of the basic components of the cell migration signalosome.

Integrins constitute a large family of 24 transmembrane receptors composed of heterodimers, each made of one α and one β subunit [for reviews see Ffrench-Constant and Colognato, 2004; Ginsberg et al., 2005; Kinashi, 2005]. In mammals, it is the combination of one of the 18 α- with one of the 8 β-subunits that sets the role of the integrin in the regulation of cell adhesion and/or migration, and confers its ligand specificity. Each type I subunit is mold on a common pattern: a large extracellular domain, a short single transmembrane domain, and a small intracellular domain. It is the cytoplasmic domain that will be connected to the actin cytoskeleton and to downstream signalling molecules such as focal adhesion kinase (FAK), integrin-linked kinase (ILK) and c-Src.

Interestingly, integrins are capable of bidirectional signalling mediating both outside-in and inside-out signals, thereby sensing and informing the cell about the external environment but also providing informations to the extracellular compartment. Integrin activity is accurately controlled by several fine mechanisms. The ligand affinity is regulated by changes in the conformation of the integrins, a process called integrin activation. Folded, intermediate and fully opened conformations exhibit increasing affinity reflecting their state of activation. These conformational changes are normally triggered by inside-out signalling initiated for instance by chemokines [Kinashi, 2005]. Integrin avidity refers to the ability of integrins to form cluster and thus to the modulation of the size of the adhesive contacts (from the smaller focal complex to mature focal adhesion up to the very large adhesion plaque) and to increasing mechanical force that can be sustained by the cell [Roca-Cusachs et al., 2009]. Outside-in signalling induced by the binding of the ligand to the integrin promotes integrin clustering and also conformational changes.

Integrins possess numerous partners that can precisely modulate their activity, adjusting affinity, avidity and signalling to exactly meet the cellular requirements [Porter, 1998; Kinashi, 2005]. The integrin interactome is rich of 156 identified components that interact with the integrins and constitute the integrin adhesome [Zaidel-Bar et al., 2007]. Members of the adhesome include FAK, ILK, c-Src, talin, integrin-associated protein (IAP), tetraspanins, and uPAR. The integrins and the adhesome form a perfect system for a motile cell combining associating informative, executive and structural functions. Thus, it is quite obvious that the advantage of associating this system with the uPAR system provide the cell with an incredible, powerful and versatile regulatory unit in charge of the migration machinery.

uPAR Structure

uPAR is the other basic constituent of the cell migration signalosome. uPAR has a specific structure.

uPAR is constituted by a 283-residue single-peptidic chain organized into three homologous domains. Domain I is located at its amino-terminus while the post-translationally-added glycosyl-phosphatidyl-inositol (GPI) anchor is present at the C-terminus of domain III [Blasi and Carmeliet, 2002; Degryse, 2003, 2008; Ragno, 2006; Archinti et al., 2010]. The three external domains form a globular-like structure creating a central pocket where uPA can bind [Llinas et al., 2005; Huang et al., 2005; Barinka et al., 2006; Huai et al., 2006]. In addition, the whole external part of uPAR constitutes a very large surface conveniently available for the binding of the other soluble ligands and lateral partners of uPAR (Table 2).

Table 2. The uPAR interactome

Type	Name of the Partner	Nature	Main Functions
Soluble External Ligand	Pro-urokinase	Zymogen	Precursor of urokinase Cell adhesion, migration and proliferation
	High molecular weight form urokinase	Serine protease	Activates plasminogen into plasmin Cell adhesion, migration and proliferation
	PAI-1/uPA complex	Complex of urokinase and its physiological inhibitor PAI-1	Proteolytically inactive Blocks uPA-dependent cell migration Promotes uPAR and integrin internalization
	Vitronectin	Extracellular matrix and plasma protein	Cell adhesion and migration Cofactor of PAI-1
	Two chain high molecular weight kininogen (HKa)	Plasma protein	Pro-inflammatory Inhibits vitronectin-dependent cell adhesion and migration
	SRPX2	Secreted protein	Development of brain speech areas
	Streptococcal surface deshydrogenase (SDH)	Anchor-less microbial surface protein from *Streptococcus Pyogenes*	Glycolytic enzyme Induces bacterial adherence to host cells
Lateral Membrane-Bound Partner	Urokinase receptor	GPI-anchored protein	Plasminogen activation Pericellular proteolysis Cell anchorage, migration and proliferation
	FPRL1, FPRL2, FPR	Seven-transmembrane domain receptors	Cell migration
	$\alpha3\beta1$, $\alpha4\beta1$, $\alpha5\beta1$, $\alpha6\beta1$, $\alpha9\beta1$, $\alpha v\beta3$, $\alpha v\beta5$, $\alpha v\beta6$, $\alpha M\beta2$, $\alpha L\beta2$, $\alpha X\beta2$	Integrins	Cell adhesion and migration
	EGFR, PDGFR, IGF-1R	Receptor tyrosine kinases	Cell proliferation and migration

Table 2. (Continued)

Type	Name of the Partner	Nature	Main Functions
	LRP-1, LRP1B, VLDL-R, mannose 6-phosphate-R, endo180	Endocytic receptors	uPAR internalization (LRP-1 induces both uPAR and integrin internalization)
	Caveolin	Scaffolding/structural protein	Intracellular signalling
	gp130	Cytokine receptor	Cell migration
	gC1qR	Complement receptor	Complement activation
	L-selectin	Adhesion receptor	Cell adhesion
	Seprase	Serine protease	Pericellular proteolysis

Several important sequences have been identified within this receptor. Domain II harbors the D2A sequence, residues $_{130}$IQEGEEGRPKDDR$_{142}$ of human uPAR, which was identified as the first region of uPAR involved in direct lateral interactions with integrins $\alpha v\beta 3$ and $\alpha 5\beta 1$ initiating integrin- but not uPAR-dependent signalling, and cell migration [Degryse et al., 2005]. Furthermore, D2A has also mitogenic activity [Degryse et al., 2007; Eden et al., in preparation]. The D2A sequence is thus the sole region of uPAR possessing intrinsic motogenic and mitogenic activities [Degryse et al., 2005; Eden et al., in preparation]. The GEEG residues constitute the minimum active sequence of D2A [Degryse et al., 2005]. Substituting the two glutamic acids for two alanines in that motif generated two inhibitors of cell migration, the D2A-Ala and GAAG peptides which interfere with uPAR-integrin interactions [Degryse et al., 2005]. Interestingly, D2A-Ala and GAAG are also inhibitors of cell proliferation and tumour growth in vivo [Eden et al., in preparation].

In the domain III of uPAR, other sites of uPAR-integrin interactions are also present [Chaurasia et al., 2006; Wei et al., 2007]. The sequence $_{240}$GCATASMCQ$_{248}$ of human uPAR is involved in the interaction with $\alpha 5\beta 1$ integrin [Chaurasia et al., 2006]. In addition, the introduction of single point mutation S245A or H249A, and D262A alters uPAR association with $\alpha 5\beta 1$ and $\alpha 3\beta 1$ respectively [Chaurasia et al., 2006; Wei et al., 2007].

Two linker regions are present within uPAR: one in between domain I and II, and another one in between domain II and III. The former DI-DII linker region contains the chemotactically active sequence $_{88}$SRSRY$_{92}$ (human sequence) [Fazioli et al., 1997]. This motif binds to the 7TMRs: FPRL1, FPRL2 and FPR resulting in the stimulation of cell migration [Resnati et al.,

1996, 2002; Fazioli et al., 1997; Degryse et al., 1999; de Paulis et al., 2004; Gargiulo et al., 2005; Selleri et al., 2005]. The mechanism of action is dependent on a change of conformation induced by uPA binding to uPAR, leading to the exposure of the SRSRY epitope. This change of conformation triggers the metamorphosis from the "receptor" stage (in which uPAR behaves as a receptor for its soluble ligand uPA) to the "ligand" stage in which uPAR binds to a 7TMR which in its turn initiates signalling and cell migration. Thus, the SRSRY motif is responsible for the migration-promoting properties implicating uPAR as a cell surface chemokine acting like the classical soluble chemokines via binding to 7TMRs [Blasi, 1999; Fazioli et al., 1997; Degryse et al., 1999]. uPAR was proposed to be classified as MACKINE (Membrane-Anchored ChemoKINE-like protein) [Degryse, 2003].

The presence of the GPI anchor gives an original touch to the structure of uPAR. However, GPI anchoring is not rare but rather widely distributed among the eukaryotic organisms including protozoa, yeasts, fungi, plants, insects, and mammals [for reviews see Orlean and Menon, 2007; Paulick and Bertozzi, 2008]. GPI biosynthesis deficiency provokes fetal lethality demonstrating that GPI anchors are critically important for the normal embryonic development of mammals [Orlean and Menon, 2007; Paulick and Bertozzi, 2008]. More than 20 genes are involved in the biosynthesis of the GPI anchor occurring in the endoplasmic reticulum. When complete, the whole anchor is bound to the protein bearing a GPI anchor sequence signal at its C-terminus. Three domains, a phosphoethanolamine linker (that is bound to the C-terminus of the protein), a conserved glycan core, and a phospholipid tail form the common structure of the GPI anchor. The glycan core is responsible of the variability among GPI anchors [Orlean and Menon, 2007; Paulick and Bertozzi, 2008]. It is thought that the nature of the GPI anchor may influence the functions of the proteins [Nicholson and Stanners, 2006]. However, the exact function of the GPI anchors is under discussion, and no correlation has been evidenced between the structure and the function of the GPI anchors [Paulick and Bertozzi, 2008].

Very little knowledge can be claimed about the function of the GPI anchor of uPAR. The GPI anchor constitutes a convenient and efficient link to the cell surface, which influences the conformation of uPAR. Soluble and GPI-bound uPAR have different conformations [Hoyer-Hansen et al., 2001; Andolfo et al., 2002]. Interestingly, the GPI anchor of Thy-1 similarly influences the conformation of Thy-1 [Barboni et al., 1995]. Since uPAR is particularly present in the lipid rafts and caveolae [Okada et al., 1995; Stahl and Mueller, 1995; Koshelnick et al., 1997; Wei et al., 1999; Schwab et al., 2001;

Cunningham et al., 2003; Sitrin et al., 2004; Sahores et al., 2008], the GPI anchor may be viewed as a signal to address uPAR in these specific domains of the plasma membrane [Varma and Mayor, 1998; Friedrichson and Kurzchalia, 1998; Nicholson and Stanners, 2006]. However, uPA binding to uPAR and the dimerization of the receptor have also been suggested to influence uPAR redistribution into lipid rafts [Sidenius et al., 2002; Cunningham et al., 2003; Sahores et al., 2008]. Last, two separate studies attributed opposite importance to the GPI anchor of uPAR [Li et al., 1994; Madsen et al., 2007]. In the former study, the GPI anchor does not interfere with uPAR functioning [Li et al., 1994]. The comparison of wild-type uPAR with a chimeric uPAR made of the three extracellular domains of uPAR attached to the transmembrane and intracellular domains of the α chain of the IL-2 receptor, revealed that the kinetics of binding, internalization and degradation of the uPA/plasminogen activator inhibitor-1 (PAI-1) complex were identical [Li et al., 1994]. These experimental data are consistent with the observation that cleaved suPAR (exposing the SRSRY epitope) can mimic the effects of uPAR suggesting that the GPI anchor play a minor role in the functioning of uPAR [Resnati et al., 1996]. In the more recent study, a chimeric receptor composed of the GPI anchor of uPAR associated with the serpin domain of PAI-1, was compared with wild-type uPAR [Madsen et al., 2007]. PAI-1 was chosen because both serpin and uPAR binds to VN in a similar region, the somatomedin B (SMB) domain [Okumura et al., 2002]. The chimeric receptor reproduced all the effects of uPAR on cell morphology suggesting that the external domains of uPAR play a minor role [Madsen et al., 2007].

Several forms of uPAR have been described which have specific binding and signalling properties [Ragno, 2006]. The DI-DII linker region is highly sensitive to proteolytic cleavage by a variety of proteases including uPA. The limited proteolytic degradation of uPAR produces DI and DIIDIII-uPAR fragments [Ploug and Ellis, 1994]. DIIDIII-uPAR can be chemotactically active reproducing the motogenic effects of uPA suggesting that the proteolytic cleavage caused the exposure of the SRSRY epitope and that DIIDIII-uPAR binds to 7TMRs [Fazioli et al., 1997]. However, uPAR can be cleaved at other sites in the same DI-DII linker region and not all the DIIDIII-uPAR fragments possess the intact exposed SRSRY motif. Furthermore, the degradation of the GPI anchor generates the soluble forms of uPAR (suPARs). High levels of the suPAR are markers of cancers and correlate with poor clinical prognosis [Stephens et al., 1999; Mustjoki et al., 1999; Sier et al., 1999]. All the DIIDIII-uPAR fragments cannot bind to uPA or VN. Only the two-chain kinin-free high molecular weight kininogen (HKa) binds to domains II and III of both full-length uPAR and DIIDIII-uPAR in a Zn^{2+}-dependent manner [Colman et al., 1997; Chavakis et al., 2000]. Interestingly, VN and HKa are competitive binding partners because both bind to domains II and III of uPAR, giving to the receptor opposite adhesive and anti-adhesive properties.

uPAR, the Migration Leader

uPAR is certainly the less expected receptor to fulfill the function of cell migration leader because uPAR is a GPI membrane-bound receptor which possesses no intracellular domain, and thus uPAR cannot directly interact with downstream signalling effectors [for reviews see Blasi and Carmeliet, 2002; Degryse, 2003, 2008; Ragno 2006; Binder et al., 2007; Mazar 2008; Archinti et al., 2010]. Nevertheless, uPAR exerts three main basic functions that plainly justify its role as cell migration leader. uPAR is known as a protease (that binds uPA), anchorage (that binds vitronectin), and signalling receptor. We now discuss these three main functions.

uPAR was originally known as the membrane receptor of urokinase (uPA), its main ligand. uPAR has two main regulatory functions, enhancing the activation rate of pro-uPA and localizing uPA activity. The pro-uPA binding to PAR increases the rate of activation of the zymogen into the fully active uPA, thereby enhancing the activation of the inactive plasminogen into the active broad-spectrum plasmin [Ellis et al., 1989]. uPA binding to uPAR also controls the location of uPA proteasic activity at discrete points on the cell surface. By controlling pericellular proteolysis, uPAR keeps the hand in the degradation of extracellular matrix (ECM) and of basement membrane proteins, a process of special importance in the context in cell migration because it permits the cell to move across the tissue [Blasi and Carmeliet, 2002]. In addition, uPA activity can produce indirect effects through the activation of growth factors such as basic fibroblast growth factor (bFGF), pro-transforming growth factor-β (pro-TGF-β), and pro-hepatocyte growth factor (pro-HGF) [Odekon et al., 1992; 1994; Naldini et al., 1992; 1995].

By binding to VN, uPAR is also a direct adhesion receptor [Walz and Chapman, 1994. Wei et al., 1994]. In addition, by regulating integrin activity, uPAR is also an indirect adhesion receptor. These functions are essential for the motile cell. Because of the locomotion, the cell is constantly renewing its adhesive contacts and there is a balance between adhesion and de-adhesion. However, if this balance gets lost, the cell may detach and anoikis is immediately there with its brutal conclusion for the cell. VN is a main constituent of the ECM. It is a common adhesive glycoprotein which is also present in plasma (at 200-400 μg/ml) and in platelets [Schvartz et al., 1999]. VN exerts crucial physiological functions in cell adhesion and migration, cell proliferation, fibrinolysis and hemostasis, and the immune response. VN is also involved in pathological processes such as thrombosis, atherosclerosis, restenosis and cancers [Schvartz et al., 1999]. VN is a main ligand of the $\alpha v\beta 3$, $\alpha IIb\beta 3$, $\alpha v\beta 5$, $\alpha v\beta 1$, $\alpha v\beta 6$ and $\alpha v\beta 8$ integrins which modulate cytoskeleton organization, downstream signalling pathways, and gene expression.

VN is also one ligand of uPAR promoting cell anchorage but apparently failing to initiate signalling through uPAR binding [Stahl and Mueller, 1997; Sidenius and Blasi, 2000; Degryse et al., 2001a]. Indeed, in the smooth muscle cells uPA and VN promote cell migration through diverse receptors (uPAR and $\alpha v\beta 3$ respectively), and activate different signalling pathways [Degryse et al., 2001a]. Moreover, both full-length VN and VN_{40-459} (a fragment without SMB domain which does not bind to uPAR) have similar chemotactic activity reinforcing the idea that VN does not induce migration via its binding to uPAR [Degryse et al., 2005]. Other reports showed that VN binding to uPAR induced cell anchorage only but no cell spreading [Stahl and Mueller, 1997; Sidenius and Blasi, 2000]. However, this matter is subject to controversy. Reports suggested that VN efficiently regulates cell morphology through binding to uPAR [Kjøller and Hall, 2001; Madsen et al., 2007]. Yet, in both studies the experimental models used were cell lines overexpressing uPAR leaving open the question of the actual effects of VN in a more physiological situation with normal level of uPAR expression [Kjøller and Hall, 2001; Madsen et al., 2007]. Last, uPAR dimerization is required for VN binding [Sidenius et al., 2002; Cunningham et al., 2003], and a more recent study showed that dimers are not the active form of uPAR [Malengo et al., 2008]. As shown by the measurement of diffusion coefficients of monomers and dimers, uPAR monomers are engaged in signalling complexes and are the active form of uPAR [Malengo et al., 2008].

On the other hand, by binding to uPAR, VN can influence pericellular proteolysis by inducing uPAR relocalization into focal adhesions which are integrin clusters. There, uPAR can conversely modulate integrin activity and connection to the cytoskeleton [Blasi and Carmeliet, 2002; Degryse, 2003,

2008]. The SMB domain (residues 1-44) of VN which contains the high-affinity binding sites for both uPAR and PAI-1, is immediately proximal to the RGD sequence (residues 45-47) binding to the integrins [Seiffert and Loskutoff, 1991; Okumura et al., 2002].

Although surprising, uPAR is really a full signalling receptor capable of regulating gene expression, cell proliferation, adhesion and migration by using the service of other membrane receptors (Table 2). uPAR engages in direct lateral interactions with RTKs such as the EGF receptor (EGFR), the PDGF receptor (PDGFR) and the IGF-1 receptor (IGF-1R); 7TMRs such as FPRL1 and FPR; endocytic receptors such as the LDL receptor-related protein (LRP-1), the very low-density lipoprotein receptor (VLDLR), the mannose 6-phosphate/IGF-II receptor (CD222, CIMPR), and Endo180 (uPARAP); caveolin; the gp130 cytokine receptor, and the integrins [Blasi and Carmeliet, 2002; Degryse, 2003, 2008; Ragno 2006; Binder et al., 2007; Mazar 2008; Archinti et al., 2010]. These interactions account for uPAR-dependent signalling. However, the molecular basis of these interactions are far from being totally understood. As stated before, only one sequence (the SRSRY epitope) has been identified to bind to the 7TMRs, FPRL1, FPRL2 and FPR [Fazioli et al., 1997; Degryse et al., 1999; de Paulis et al., 2004; Gargiulo et al., 2005; Selleri et al., 2005], but so far, no uPAR-binding sequence has been reported in these 7TMRs. Other sequences of uPAR (D2A and the GCATASMCQ motif) are involved in the interactions with the integrins [Degryse et al., 2005; Chaurasia, 2006; Wei, 2007], and in the integrins, sequences interacting with uPAR have also been identified (discussed below).

Like other GPI-bound receptors, uPAR can redistribute in special micro-domains of the membrane, the lipids rafts and caveolae which are considered as convenient signalling platforms concentrating together ligands, receptors and cytoplasmic signalling effectors [Chapman et al., 1999; Wei et al., 1999; Simons and Toomre, 2000; Sidenius et al., 2002; Cunningham et al., 2003]. Dimerization of uPAR is required for VN binding and redistribution into the lipid rafts [Sidenius et al., 2002; Cunningham et al., 2003]. As a GPI-anchored protein, uPAR is supposed to have more mobility onto the plasma membrane, and in fact, uPAR mobility seems to be closely associated with its functioning. Active uPAR monomers engage in multiprotein transmembrane signalling complexes forming very large signalling complexes on particular places of the cell surface [Bohuslav et al., 1995; Resnati et al., 1996; Degryse et al., 1999; Sitrin et al., 2000; Malengo et al., 2008]. During cell migration, uPAR is redistributed at the leading edge of the migrating cell [Degryse et al., 1999].

Endocytosis is an essential part of signalling [Ceresa and Schmid, 2000; Sadowski et al., 2009]. It is also the mechanism that keeps running the uPAR system allowing to recycle back free uPAR at specific locations of the cell surface such a new focal complexes at the leading edge of the motile cell [Prager et al., 2004]. Experimental evidence supports that view and in addition three distinct mechanisms of uPAR internalization have been reported. uPAR endocytosis is a convenient and powerful way of controlling both uPAR and integrin activities and their location [Degryse et al., 2001b; Czekay et al., 2003]. Cell migration stops when uPAR regeneration on the cell surface is discontinued, [Degryse et al., 2001b]. On the other hand, the blockade of uPAR internalization enhances uPA/uPAR synthesis and expression on the cell surface, leading to increased plasminogen activation, cell motility and tumoural cell invasion [Weaver et al., 1997; Webb et al., 1999, 2000; Sid et al., 2006].

The physiological inhibitor of uPA, PAI-1 has a central role in uPAR internalization, which is actually started up by this serpin. The PAI-1/uPA complex interacts with the endocytic receptor LRP-1 and in a latter step permits direct interaction of uPAR with LRP-1 [Nykjaer et al., 1992, 1994; Conese et al., 1995; Czekay et al., 2001]. The LRP-1/PAI-1/uPA/uPAR complex is then internalized via the clathrin-coated pathway, and while free uPAR will be recycled back to the cell surface from the early endosomes the PAI-1/uPA complex will be degraded [Nykjaer et al., 1997]. In this endocytic mechanism, an intact uPAR (for its uPA binding capacity) is required. The shorter form DIIDIII-uPAR is poorly internalized because uPA, uPA/PAI-1 and LRP-1 do not bind to DIIDIII-uPAR, and consequently this shorter form of uPAR is not efficiently internalized [Høyer-Hansen et al., 1992; Nykjaer et al., 1998; Ragno et al., 1998]. This fact may explain why tumour cells exhibit more copies of DIIDIII-uPAR than of full-length uPAR on their cell surface [Ragno et al., 1998]. By inducing the internalization of uPAR and integrins, PAI-1 inhibits uPA-induced cell migration and promotes cell detachment [Degryse et al., 2001b; Czekay et al., 2003]. Thus, besides controlling the uPAR system, uPAR internalization serves modulating the activity and the number of integrins present on the cell surface, which is extremely precious in the perspective of cell migration [Czekay et al., 2003].

Another mechanism of uPAR internalization requires the mannose 6-phosphate/insulin-like growth factor-II receptor (Cation-Independent Mannose 6-Phosphate Receptor, CIMPR), which negatively regulates uPAR functions [Leksa et al., 2002]. The CIMPR-dependent route is unique because it internalizes both uPAR and DIIDIII-uPAR in an uPA-independent manner [Nykjaer et al., 1998].

The last known mechanism of internalization and recycling of uPAR has been recently discovered [Cortese et al., 2008]. A PAI-1/uPA- and lipid rafts-independent micropinocytic-like route consents the endocytosis and rapid recycling of uPAR [Cortese et al., 2008]. These data further enlighten the importance of regenerating free uPAR on the cell surface.

The Molecular Interactions between uPAR and the Integrins

The exact mechanism driving the interactions between uPAR and the integrins is unknown, however we do have some clues about the molecular interactions between uPAR and the integrins. We know that uPA triggers these interactions, positively or negatively influencing most if not all uPAR-integrin interactions. Due to this uPA function, it is thought that the domain I of uPAR, essential for uPA binding, is importantly contributing either directly or indirectly to the formation of uPAR-integrin interactions [Myöhänen et al., 1993; Montuori et al., 2002]. The fact that the shorter form of uPAR, DIIDIII-uPAR does not bind to uPA and to the integrins further supports that idea [Ragno et al., 1998; Montuori et al., 1999, 2002]. Nevertheless, none of the sequences interacting with the integrins have been identified in the domain I, rather they were located in the domains II and III [Degryse et al., 2005; Chaurasia et al., 2006; Wei et al., 2007]. Moreover, sequences involved in the uPAR-integrin interactions have also been reported in the integrins [for a recent review see Degryse, 2008].

The D2A sequence $_{130}$IQEGEEGRPKDDR$_{142}$ (human uPAR) harbored by domain II is involved in uPAR-integrin interactions [Degryse et al., 2005]. D2A interacts specifically with $\alpha v\beta 3$ and $\alpha 5\beta 1$ integrins inducing integrin-dependent signalling and cell migration. Replacing the two glutamic acids into two alanines abolished the chemotactic activity of D2A. The derived synthetic peptides D2A-Ala and GAAG are potent inhibitors of cell migration acting by disrupting the uPAR association with the integrins. Moreover, we recently evidenced that both D2A-Ala and GAAG peptides were also inhibitors of cell proliferation, and that the sequence D2A was mitogenic representing the first ever identified mitogenic sequence of uPAR [Eden et al., in preparation]. These data agree with previous reports of the literature demonstrating that integrins regulates RTK activity and thus cell growth and survival [Miyamoto et al., 1996; Danilkovitch-Miagkova et al., 2000; Streuli and Akhtar, 2009]. Therefore, by promoting/modulating the cooperation between integrins and RTKs, uPAR can form larger signalling complexes, i.e. uPAR proliferasomes, that are implicated in cell proliferation and may explain some of the functions of uPAR in pathologies such tumour growth. In our study published in 2005, we hypothesized that other site(s) of interaction between uPAR and integrins were likely to be present in domain III [Degryse et al., 2005]. Two observations supported this view. First, the chemotaxis effect of D2A was dependent on uPAR expression by the cells. Second, cells expressing the short DIDII-uPAR were less sensitive to very low doses of VN [Degryse et al., 2005].

Two separate studies validated our hypothesis reporting other sites of interaction with the integrins within the domain III of uPAR [Chaurasia et al., 2006; Wei et al., 2007]. The motif corresponding to $_{240}$GCATASMCQ$_{248}$ of human uPAR is implicated in the interaction with α5β1 integrin [Chaurasia et al., 2006]. However, the derived synthetic peptide GCATASMCQ did not promote cell signalling, and rather reduced integrin-dependent ERK activation [Chaurasia et al., 2006]. In addition, single point mutation of residue S245A or H249A altered the association of uPAR with the same α5β1 integrin while mutation D262A disrupted the interaction with α3β1 integrin [Chaurasia et al., 2006; Wei et al., 2007]. Furthermore, mutating uPAR at both H249A and D262A disrupted all the interactions with α5β1 and α3β1 integrins but neither affected the lateral interactions with caveolin, EGFR and αv integrins, nor altered the binding to uPA and VN. However, if uPA binding was preserved, signalling was reduced along with cell invasion and tumour growth in vivo, further enlightening the biological importance of uPAR-integrin interactions [Tang CH et al., 2008]. In summary, uPAR harbors at least two integrin sites, one is located in the domain II and another one in the domain III [Degryse et al., 2005; Chaurasia et al., 2006; Wei et al., 2007]. Only the D2A motif from domain II has intrinsic motogenic and mitogenic activities [Degryse et al., 2005, 2007; Eden et al., in preparation].

Within the integrins, the sequences responsible of the uPAR-integrin association have been identified on both subunits. In the β-propeller of the αM subunit of αMβ2 (Mac-1, CR3, CD11b/CD18) integrin, the sequence $_{424}$PRYQHIGLVAMFRQNTG$_{440}$ constitutes a non-I-domain binding site for uPAR [Simon et al., 2000]. The derived synthetic M25 peptide impedes the formation of uPAR-αMβ2 and uPAR-β1 complexes [Simon et al., 2000]. Although this peptide did not inhibit the binding of ligand to αM, M25 nevertheless altered the function of the integrin blocking the adhesion of leukocytes to fibrinogen or VN and the stimulation of endothelial cells by cytokine [Simon et al., 2000]. Moreover, peptide M25 also inhibited integrin-dependent migration of SMC on FN and collagen [Simon et al., 2000]. M25 was identified by homology with the former peptide 25 with sequence STYHHLSLGYMYTLN, which came out from a phage display library and was also capable to bind to uPAR [Wei et al., 1996]. Since then, α325 another peptide matching M25 but derived from α3β1 integrin has been used to disrupt uPAR-α3β1 interaction preventing uPA synthesis, cell migration and invasion, and EGFR activation [Wei et al., 2001; Ghosh et al., 2006; Mazzieri et al., 2006].

Two other uPAR binding sequences $_{224}$NLDSPEGGF$_{232}$ and $_{262}$FHFA-GDGKL$_{270}$ are located in the β1 subunit of α5β1 [Wei et al., 2005]. These sequences are close to the RGD binding site of α5β1 and to the β-propeller domain of the α5 subunit [Wei et al., 2005]. The identification of these sequences improved our understanding of the uPAR-integrin interactions. The binding of uPAR to the β1 subunit of α5β1 has profound effects on both integrin conformation and activity but in the absence of uPAR, the integrins keep up functioning. β1P1 and β1P2, the respective peptides derived from the above β1 sequences prevented the formation of uPAR-α5β1 complexes but failed to completely abolish integrin activity. [Wei et al., 2005]. In addition, uPAR-associated α5β1 bound FN both in a RGD- and β1 peptide-dependent manner whereas free α5β1 bound to FN in RGD-dependent manner only. uPAR binding to α5β1 induced a strong cell adhesion to FN, while cells adhered weakly to FN when uPAR expression was downregulated [Wei et al., 2005].

The identifications of uPAR-binding sites in both α and β subunits of the integrins suggest that uPAR may connect to the integrins in different manners and thus the interaction of uPAR with either one or the other subunit may have different functions. Moreover, the existence of these sites of interactions on uPAR and the integrins is certainly in favour of direct uPAR-integrin association. A long list of experimental reports using distinct experimental methods comprising co-immunoprecipitation and FRET [Myöhänen et al., 1993; Xue et al., 1994, 1997; Wei et al., 1996; Kindzelskii et al., 1997] support this idea (see below). A recent report showed that it is possible to finely discriminate uPAR interactions [Tang CH et al., 2008]. uPAR-β1 interactions were destroyed by introducing the double mutations H249A-D262A in uPAR but all other connections were preserved including binding to uPA and VN, and lateral interactions with caveolin, EGFR and αv integrins [Tang CH et al., 2008].

More than Simply Adding uPAR Plus the Integrins

The cell migration signalosome is much more than simply adding the uPAR system plus the integrin adhesome. It is really a new entity with specific characteristics reflecting the influence that uPAR and the integrins exert on each other but also the synergy that develops between these players.

uPAR enters in cis-interactions with almost half of the integrin family including αMβ2 (Mac-1, CR3), αLβ2 (LFA-1), αXβ2 (CR4), α3β1, α4β1, α5β1, α6β1, α9β1, αvβ3, αvβ5, αvβ6. In addition, trans-interactions with the integrins expressed by adjacent cells suggested that uPAR connection with the integrins is involved in cell-cell contacts [Tarui et al., 2001a].

The study of the relationship between uPAR and the integrins has been the topic of numerous reports explaining why uPAR-integrin interactions are the best known among all the lateral interactions of uPAR. A brief survey of the literature reveals that the most popular experimental methods used to demonstrate uPAR-integrin interactions were mainly based on biochemical and microscopy techniques comprising co-immunoprecipitation, in vitro pull-down assay using purified proteins, colocalization, cocapping, fluorescence resonance energy transfer (FRET) [Pöllänen et al., 1988; Xue et al., 1994, 1997; Bohuslav et al., 1995; Reinartz et al., 1995; Kindzelskii et al., 1997; Ghosh et al., 2000; Wei et al., 2001; Xia et al., 2002; Gellert et al., 2004; Bass et al., 2005; Degryse et al., 2005]. Among these methods, FRET is particularly interesting because this technique is carried out in living cells, and gives precious indications about the association of uPAR with the β1 or β3 integrins which reside at a distance of 7 nm from each other on the cell surface. A distance compatibles with direct lateral interactions between uPAR and β1 or β3 integrins [Xue et al., 1997]. Finally, the finding of the sequences of both proteins involved in uPAR-interactions confirmed all these data [Wei et al., 1996, 2001, 2005, 2007; Simon et al., 2000; Ghosh et al., 2006; Mazzieri et al., 2006; Degryse et al., 2005; Chaurasia et al., 2006; Tang CH et al., 2008].

uPAR has a sharp influence on integrin that can be finely tuned to either positively or negatively regulate the activity if the integrins. In most situations, a positive regulation of integrin activity was mentioned, uPAR stimulates integrin-dependent cell adhesion and migration [Sitrin et al., 1996; Yebra et al., 1996; May et al., 1998, 2000; Degryse et al., 2005; Wei et al., 2005]. But some negative effects were also reported [Sitrin et al., 1996; Simon et al., 1996].

uPAR exerts its effects using several mechanisms that can be combined in different manners. uPAR (like other integrin regulators) can regulate integrin function in the classical way affecting either affinity, avidity, or the spatial distribution of integrins on the cell surface. Moreover, uPAR is capable of modulating the number of integrins present on the cell surface by acting on integrin expression and/or internalization. Last, uPAR also promotes the limited proteolytic degradation of the integrins which was recently showed to regulated integrin activity. Therefore, uPAR has all the required functionality to manage the work, number and location of the integrins.

Depending on the type of the integrin, uPAR is either a cofactor absolutely required for the normal functioning of the integrin or a modulator of integrin affinity. The β2 integrins αMβ2 and αLβ2 are inactive in uPAR knock-out mice. In these animals, neutrophils recruitment was deficient due to altered leukocyte adhesion to the endothelial wall [May et al., 1998; Simon et al., 2000]. In wild-type leukocytes, β2 integrin-dependent cell adhesion is lost after removal of uPAR from the cell surface [May et al., 1998]. It is difficult to understand such a drastic effect of uPAR on these β2 integrins because uPAR modulates the conformation of the external domains and orientation of the transmembrane domain of αMβ2 [Tang ML et al., 2008]. In the first instance, these effects do not appear really different than those exerted on α5β1 conformation [Wei et al., 2005]. However, α5β1 was still functioning even when not connected to uPAR [Wei et al., 2005]. The effects on β2 integrins is however in line with the action anti-uPAR antibodies which by inhibiting the activity of uPAR completely blocked αvβ3 activity preventing VN-induced rat SMC migration [Degryse et al., 1999].

uPAR controls integrin avidity by promoting integrins clustering [Myöhänen et al., 1993; Wei et al., 1996; Degryse et al., 1999; Gellert et al., 2004]. This receptor also regulates the location of the integrins onto the plasma membrane [Ghosh et al., 2000]. This process seems to be dependent on uPA (or pro-uPA) as uPA induces the co-localization of β2 integrins and uPAR in leukocytes [Bohuslav et al., 1995; Petty et al., 1997], and pro-uPA the co-localization of αvβ3 and uPAR at the leading edge of migrating SMC [Degryse et al., 1999]. VN, the ligand of uPAR and integrins, exerts a similar influence by inducing the clustering of uPAR and αvβ3 [Ciambrone et al., 1992; Xue et al., 1997; Stepanova et al., 2002]. Moreover, uPAR dispatches integrins into lipid rafts favouring the formation and stabilizing uPAR-caveolin-β1 integrin complexes [Stahl and Mueller, 1995; Chapman et al., 1999; Wei et al., 1999; Schwab et al., 2001].

uPAR-integrin interactions are dependent on the level of expression of uPAR and the integrins. In fact, uPAR expression controls the level of integrin expression, and conversely integrins rule the level uPAR expression [Bianchi et al., 1996; Wang et al., 1998; Ghosh et al., 2000; Adachi et al., 2001; Hapke et al., 2001a, b]. For example, the levels of uPAR and αvβ3 are strictly correlated [Nip et al., 1995; Adachi et al., 2001; Khatib et al., 2001]. In addition, the level of uPAR expression correlates with the activation state of the integrins [Chintala et al., 1997; Aguirre Ghiso et al., 1999; Simon et al., 2000]. Upregulating uPAR enhances the formation of uPAR-integrin complex and corresponds to an activation of the αMβ2, αLβ2 and α5β1 integrins [Aguirre-Ghiso et al., 2001; May et al., 2002]. Once again, uPA binding to uPAR thereby the conformation of this receptor may be essential, as suggested by uPAR occupancy by uPA that increases αvβ5 expression and activity [Silvestri et al., 2002]. On the other hand, the downregulation of uPAR disrupts the uPAR-integrin complex and deactivates the αMβ2, αvβ3 and α5β1 integrins [Sitrin et al., 1996; Aguirre Ghiso et al., 1999; Adachi et al., 2001; Gondi et al., 2006]. Last, uPAR expression also participates in the control of integrin localization, the level of uPAR expression correlates with the formation of focal adhesions which are clusters of integrins [Chintala et al., 1997; Kjøller and Hall, 2001; Abu-Ali et al., 2005].

The activity and number of integrins displayed on the cell membrane are also regulated by uPAR endocytosis [Czekay et al., 2003]. PAI-1 triggers uPA-uPAR-integrins complexes internalization promoting integrins deactivation and removal from the cell surface followed by cell detachment from VN, FN and collagen 1 matrices suggesting that diverse integrins are effectively endocytosed via this mechanism [Czekay et al., 2003]. Therefore, uPAR internalization is an effective way to regulate integrin activity that can be conveniently carried out to adapt the cell machinery to the various molecular requirements of the motile cells. For example, this mechanism may likely to serve disassembling the focal adhesions during cell migration.

As a protease, uPA is capable of cutting off the β-propeller domain of α6 integrins [Demetriou et al., 2004; Demetriou and Cress, 2004]. The removal of this β-propeller domain seems to boost the activity of integrins. In prostate tumoural cells, overexpression of wild-type α6 integrins resulted in a three-fold increases in cell migration on laminin when compared to the same cells expressing a non-cleavable variant of α6 [Pawar et al., 2007]. The clipping of α6 integrins seems to occur in invasive tumoural cells only and might be related to the appearance of the more aggressive invasive phenotype of cancer cells [Pawar et al., 2007].

The influence of the shorter forms of uPAR on integrin activity has not been thoroughly studied. These shorter forms do not appear to bind to the integrins [Ragno et al., 1998; Montuori et al., 1999, 2002]. However, if not associating with the integrins the shorter forms of uPAR can still exert indirect effects as shown by the DIIDIII-uPAR$_{88\text{-}274}$ (still possessing the SRSRY chemotactic sequence and binding to FPRL1) fragment which inhibits integrin-dependent adhesion by blocking chemokine-induced inside-out signalling [Furlan et al., 2004].

For their part, integrins are not passive but rather affect uPAR expression, distribution and activity. In T lymphocytes, uPAR expression is dependent on β1 and β2 integrins [Bianchi et al., 1996]. There is also a parallel between the levels of expression of αvβ3 or α6β1 and uPAR that correlates with the functioning of uPAR-integrin complex as a migration signalosome [Nip et al., 1995; Adachi et al., 2001; Khatib et al., 2001; Sawai et al., 2006]. Both uPAR- and αvβ3-dependent cell migration were blocked using anti-uPAR or anti-αvβ3 antibodies [Degryse et al., 1999, 2001a]. Similarly, antibodies against α6 and β1 integrins and uPAR reduced cell proliferation, adhesion and migration of pancreatic cancer cells [Sawai et al., 2006]. uPAR-integrin interactions are positively regulated by the ligands of the integrins. FN, VN and laminin promote the interaction of uPAR with β1, β3, α3, α5, α6, and αv integrins [Xue et al., 1997; Salasznyk et al., 2007]. In addition, integrins influence uPAR distribution on the cell surface. The aggregation of αMβ2 or α3β1 integrin re-distributes uPAR into integrin clusters [Xue et al., 1994; Ghosh et al., 2000, 2006]. Downregulating partially or completely the β2 integrins (αLβ2, αMβ2, αXβ2, αDβ2) decreased the capping of uPAR [Kindzelskii et al., 1994]. As uPAR partners, the integrins have a notable impact on uPAR activity but can also use uPAR as a tool to control the activity of other integrins. This latter property appears of appealing interest for the motile cell because the signal regulating the activity of the other integrins does not have to enter into the cell, the messenger uPAR resides onto the plasma membrane insuring a fast and convenient response. αMβ2 and α4β1 (VLA-4) integrins can activate uPAR [Wong et al., 1996; May et al., 2000]. By exploiting uPAR as a molecular mediator, α4β1 succeeds in turning on β2 integrins, and $\alpha3\beta1$ promotes the activity of other $\beta1$ integrins [May et al., 2000; Wei et al., 2001].

Both main ligands of uPAR, uPA and VN play their own part in the creation of uPAR-integrin interactions. VN has a particular place because it is a common ligand of uPAR and integrins which promotes uPAR-integrin association by acting on both sides of the complexes [Xue et al., 1997; Salasznyk et al., 2007]. However, it seems that uPA might play a similar role as well. On the one hand, $\beta2$, $\alpha5\beta1$ and $\alpha v\beta5$ integrins require uPA binding to uPAR to function normally suggesting that these interactions are dependent on uPAR conformation [Simon et al., 1996; Yebra et al., 1996, 1999; Carriero et al., 1999; Chavakis et al., 1999; Wei et al., 2001; Silvestri et al., 2002; Degryse et al., 2005; Margheri et al., 2006]. The fact that suPAR associates poorly with the integrins further supports this idea [Degryse et al., 2005]. On the other hands, several reports demonstrated that uPA actually binds to the integrins [Pluskota et al., 2003, 2004; Demetriou et al., 2004; Kwak et al., 2005; Franco et al., 2006; Tarui et al., 2006]. In the bridge model attempting to explain uPAR-integrin interactions, uPA brings uPAR and the integrin together thereby initiating their interactions and subsequent signalling [for a review see Degryse, 2008]. This model is compatible with the structure of uPA which binds to uPAR through its growth factor domain (GFD) and can simultaneously interacts with the integrin via its kringle domain or via its catalytic domain [Plustoka et al., 2003, 2004; Demetriou et al., 2004; Kwak et al., 2005; Franco et al., 2006; Tarui et al., 2006; Pawar et al., 2007]. In that way, the GFD of a single uPA binds to uPAR while the kringle or proteolytic domain of the same uPA connects to the I-domain of $\alpha M\beta2$ integrin promoting cell adhesion and migration, and enhancing plasminogen activation and fibrinolysis [Pluskota et al., 2003, 2004]. Evidence showed that the uPA kringle binds to $\alpha v\beta3$, $\alpha IIb\beta3$, $\alpha4\beta1$ and $\alpha9\beta1$ integrins [Kwak et al., 2005; Tarui et al., 2006; Degryse et al., 2008]. The uPA kringle binding to $\alpha v\beta3$ actually stimulates integrin activity kicking off an array of signalling effectors such as $G_{i/o}$ protein, PI-3 kinase, ERK and p38 MAP kinases, and the EGF-R, and inducing expression of cytokines, cell adhesion and migration, and plasminogen activation [Kwak et al., 2005; Tarui et al., 2006]. However, it has to be said that the binding of the isolated uPA kringle domain does not necessarily reproduce the effects of full-length uPA. In contrast to uPA but like other kringle fragments such as angiostatin (K1-4) and fragments K1-3 and K1-5 of plasminogen, the uPA kringle inhibits angiogenesis and tumour growth [Tarui et al., 2001b; Kim KS et al., 2003a, b; Kim CK et al., 2007].

After having summarized the intricated relationship between uPAR and the integrins, not forgetting the influence of their ligands, it is time to discuss the advantage of combining the uPAR system with the integrin adhesome in order to generate the cell migration signalosome.

By associating with the integrins, uPAR gains access to a large family of potential transducers capable of mediating uPAR-dependent signals to the cell, and in charge of mobilizing the cell machinery comprising the cytoskeleton. In some way, the formation of the uPAR-integrin complex connects the head with the shoes of cell migration. But there is more, uPAR becomes a biomechanical sensor and mediator of biomechanical forces thereby acquiring a new essential function for the regulation of cell migration. As biomechanical sensor, uPAR transmits the external forces to the actin cytoskeleton and translates these forces into biochemical signals [Wang et al., 1995]. In particular, external forces promote the formation of focal complexes that mature into focal adhesions [Katsumi et al., 2004, 2005].

When connecting to the integrins, uPAR may modify the binding capacity of the integrin for its normal ligands. This uPAR-dependent forced change in ligand binding preference has been observed for the β1 integrins. In uPAR-transfected HEK 293 cells, uPAR shifted β1 integrin-dependent cell adhesion to FN into VN [Simon et al., 1996; Wei et al., 1996]. This uPAR-forced change in ligand preference constitutes a very simple adaptive mechanism for the motile cell providing easy cellular adjustment to diverse matrices.

uPAR binding to integrins boosts integrin-dependent signalling and migration [Simon et al., 1996; Chavakis et al., 1999; Yebra et al., 1999; Degryse et al., 2005]. Through the integrins, uPAR affects FAK, the MAP kinase pathway (Ras, MEK and ERK), MLCK (myosin light chain kinase), caveolin, Jaks (JAnus Kinases) and Stats (Signal Transducer and Activator of Transcription) [Simon et al., 1996; Koshelnick et al., 1997; Dumler et al., 1998, 1999a, b; Chavakis et al., 1999; Nguyen et al., 1999; Wei et al., 1999; Yebra et al., 1999; Degryse et al., 2005]. c-Src tyrosine kinase activity is required for uPAR-dependent cell migration, and seems to rely on the formation of the uPAR-integrin complex and on the interaction of c-Src with the integrins [Degryse et al., 1999; Wei et al., 1999].

On the one hand, the uPAR system brings together proteolytic, anchorage and signalling activities. This system plays a critical role in cell proliferation, adhesion and migration, angiogenesis, pericellular proteolysis, plasminogen and growth factors activation, tissue remodeling, tumour development and invasion and metastasis. On the other hand, the integrins form a large family of membrane receptors that mediate bi-directional signalling and are connected to the cell cytoskeleton. Integrins are also essential partners of growth factor and cytokine receptors, and the integrin adhesome is exceptionally rich of integrin partners. However, new partners may be brought by the combination of the adhesome with the uPAR interactome. The integrins exerts vital functions in embryonic development and in adult tissue homeostasis, inflammation, cell differentiation, migration, adhesion and survival, cell cycle progression, angiogenesis and tumour metastasis, and confer resistance to genotoxic injury i.e. adhesion-mediated radioresistance/ drug resistance of tumour cells. Considering these unique properties, it seems quite obvious that concentrating all of these characteristics into one single regulatory unit i.e. the cell migration signalosome gives to the motile cell an ultimate advantage.

Conclusion

Merging the uPAR system that allow the motile cell to move across and modify its environment with the integrin adhesome that consent to the same cell to adapt to its new environment and survive, benefits greatly to the migrating cell. The uPAR-integrin complex constitutes the basic building block of the cell migration signalosome on the top of which the other components of the signalosome will be added in order to satisfy the specific requirements of the motile cell in the diverse crossed environments throughout its journey. Thus, the cell migration signalosome represents some kind of "chameleon signalosome" i.e. a versatile signalling complex. Targeting this signalosome offers a unique opportunity to inhibit diverse receptors and downstream signalling pathways, and provide a new strategy to overcome pathologies such as cancers in which cell migration plays a major role. Our most recent inhibitors D2A-Ala and GAAG, may be the prototype of future anti-tumoural drugs [Degryse et al. 2005].

References

Abu-Ali, S; Sugiura, T; Takahashi, M; Shiratsuchi, T; Ikari, T; Seki, K; Hiraki, A; Matsuki, R; Shirasuna, K. Expression of the urokinase receptor regulates focal adhesion assembly and cell migration in adenoid cystic carcinoma cells. *J Cell Physiol,* 2005, May; 203(2), 410-419.

Adachi, Y; Lakka, SS; Chandrasekar, N; Yanamandra, N; Gondi, CS; Mohanam, S; Dinh, DH; Olivero, WC; Gujrati, M; Tamiya, T; Ohmoto, T; Kouraklis, G; Aggarwal, B; Rao, JS. Down-regulation of integrin alpha(v)beta(3) expression and integrin-mediated signalling in glioma cells by adenovirus-mediated transfer of antisense urokinase-type plasminogen activator receptor (uPAR) and sense p16 genes. *J Biol Chem,* 2001, Dec 14, 276(50), 47171-47177.

Aguirre Ghiso, JA; Kovalski, K; Ossowski, L. Tumor dormancy induced by downregulation of urokinase receptor in human carcinoma involves integrin and MAPK signalling. *J Cell Biol,* 1999, Oct 4, 147(1), 89-104.

Aguirre-Ghiso, JA; Liu, D; Mignatti, A; Kovalski, K; Ossowski, L. Urokinase receptor and fibronectin regulate the ERK(MAPK) to p38(MAPK) activity ratios that determine carcinoma cell proliferation or dormancy in vivo. *Mol Biol Cell,* 2001, Apr, 12(4), 863-879.

Albiges-Rizo, C; Destaing, O; Fourcade, B; Planus, E; Block, MR. Actin machinery and mechanosensitivity in invadopodia, podosomes and focal adhesions. *J Cell Sci,* 2009, Sep 1, 122(Pt 17), 3037-3049.

Andolfo, A; English, WR; Resnati, M; Murphy, G; Blasi, F; Sidenius, N. Metalloproteases cleave the urokinase-type plasminogen activator receptor in the D1-D2 linker region and expose epitopes not present in the intact soluble receptor. *Thromb Haemost,* 2002, Aug, 88(2), 298-306.

Archinti, M; Eden, G; Murphy, R; Degryse, B. Follow the leader: when the urokinase receptor coordinates cell adhesion, motility and proliferation with cytoskeleton organization. In: Sébastien Lansing, Tristan Rousseau editors. *Cytoskeleton: Cell Movement, Cytokinesis and Organelles Organization*. New York: Nova Publishers, 2010, In the press.

Aubert, M; Badoual, M; Christov, C; Grammaticos, B. A model for glioma cell migration on collagen and astrocytes. *J R Soc Interface,* 2008, Jan 6, 5(18), 75-83.

Bagorda, A; Mihaylov, VA; Parent, CA. Chemotaxis: moving forward and holding on to the past. *Thromb Haemost,* 2006, Jan, 95(1), 12-21.

Barboni, E; Rivero, BP; George, AJ; Martin, SR; Renoup, DV; Hounsell, EF; Barber, PC; Morris, RJ. The glycophosphatidylinositol anchor affects the conformation of Thy-1 protein. *J Cell Sci,* 1995, Feb, 108(Pt 2), 487-497.

Barinka, C; Parry, G; Callahan, J; Shaw, DE; Kuo, A; Bdeir, K; Cines, DB; Mazar, A; Lubkowski, J. Structural basis of interaction between urokinase-type plasminogen activator and its receptor. *J Mol Biol,* 2006, Oct 20, 363(2), 482-495.

Bass, R; Werner, F; Odintsova, E; Sugiura, T; Berditchevski, F; Ellis, V. Regulation of urokinase receptor proteolytic function by the tetraspanin CD82. *J Biol Chem,* 2005, Apr 15, 280(15), 14811-14818.

Bianchi, E; Ferrero, E; Fazioli, F; Mangili, F; Wang, J; Bender, JR; Blasi, F; Pardi, R. Integrin-dependent induction of functional urokinase receptors in primary T lymphocytes. *J Clin Invest,* 1996, Sep 1, 98(5), 1133-1141.

Binder, BR; Mihaly, J; Prager, GW. uPAR-uPA-PAI-1 interactions and signalling: a vascular biologist's view. *Thromb Haemost,* 2007, Mar, 97(3), 336-342.

Blasi, F. The urokinase receptor. A cell surface, regulated chemokine. *APMIS,* 1999, Jan;107(1), 96-101.

Blasi, F; Carmeliet P. uPAR: a versatile signalling orchestrator. *Nat Rev Mol Cell Biol,* 2002, Dec;3(12), 932-943.

Bohuslav, J; Horejsí, V; Hansmann, C; Stöckl, J; Weidle, UH; Majdic, O; Bartke, I; Knapp, W; Stockinger, H. Urokinase plasminogen activator receptor, beta 2-integrins, and Src-kinases within a single receptor complex of human monocytes. *J Exp Med,* 1995, Apr 1, 181(4), 1381-1390.

Bretscher, MS. On the shape of migrating cells--a 'front-to-back' model. *J Cell Sci,* 2008, Aug 15, 121(Pt16), 2625-2628.

Burke, B; Roux, KJ. Nuclei take a position: managing nuclear location. *Dev Cell,* 2009, Nov;17(5), 587-597.

Carriero, MV; Del Vecchio, S; Capozzoli, M; Franco, P; Fontana, L; Zannetti, A; Botti, G; D'Aiuto, G; Salvatore, M; Stoppelli, MP. Urokinase receptor interacts with alpha(v)beta5 vitronectin receptor, promoting urokinase-dependent cell migration in breast cancer. *Cancer Res,* 1999, Oct 15, 59(20), 5307-5314.

Caton, R. Contributions to the cell-migration theory. *J Anat Physiol,* 1870, 5, 35-420.7.

Ceresa, BP; Schmid, SL. Regulation of signal transduction by endocytosis. *Curr Opin Cell Biol,* 2000, Apr; 12(2), 204-210.

Chapman, HA; Wei, Y; Simon, DI; Waltz, DA. Role of urokinase receptor and caveolin in regulation of integrin signalling. *Thromb Haemost,* 1999, Aug;82(2), 291-297.

Charest, PG; Firtel, RA. Big roles for small GTPases in the control of directed cell movement. *Biochem J,* 2007, Jan 15, 401(2), 377-390.

Chaurasia, P; Aguirre-Ghiso, JA; Liang, OD; Gardsvoll, H; Ploug, M; Ossowski, L. A region in urokinase plasminogen receptor domain III controlling a functional association with alpha5beta1 integrin and tumor growth. *J Biol Chem,* 2006, May 26, 281(21), 14852-14863.

Chavakis, T; May, AE; Preissner, KT; Kanse, SM. Molecular mechanisms of zinc-dependent leukocyte adhesion involving the urokinase receptor and beta2-integrins. *Blood,* 1999, May 1, 93(9), 2976-2983.

Chavakis, T; Kanse, SM; Lupu, F; Hammes, HP; Müller-Esterl, W; Pixley, RA; Colman, RW; Preissner, KT.Different mechanisms define the antiadhesive function of high molecular weight kininogen in integrin- and urokinase receptor-dependent interactions. *Blood,* 2000, Jul 15, 96(2), 514-522.

Chen, CS. Mechanotransduction - a field pulling together? *J Cell Sci* 2008 Oct, 15, 121(Pt20), 3285-3292.

Chintala, SK; Mohanam, S; Go, Y; Venkaiah, B; Sawaya, R; Gokaslan, ZL; Rao, JS. Altered in vitro spreading and cytoskeletal organization in human glioma cells by downregulation of urokinase receptor. *Mol Carcinog,* 1997, Dec, 20(4), 355-365.

Ciambrone, GJ; McKeown-Longo, PJ. Vitronectin regulates the synthesis and localization of urokinase-type plasminogen activator in HT-1080 cells. *J Biol Chem,* 1992, Jul 5, 267(19), 13617-13622.

Colman, RW; Pixley, RA; Najamunnisa, S; Yan, W; Wang, J; Mazar, A; McCrae, KR. Binding of high molecular weight kininogen to human endothelial cells is mediated via a site within domains 2 and 3 of the urokinase receptor. *J Clin Invest,* 1997, Sep, 15, 100(6), 1481-1487.

Conese, M; Nykjaer, A; Petersen, CM; Cremona, O; Pardi, R; Andreasen, PA; Gliemann, J; Christensen, EI; Blasi, F. alpha-2 Macroglobulin receptor/Ldl receptor-related protein(Lrp)-dependent internalization of the urokinase receptor. *J Cell Biol,* 1995, Dec, 131(6Pt1), 1609-1622.

Cortese, K; Sahores, M; Madsen, CD; Tacchetti, C; Blasi, F. Clathrin and LRP-1-independent constitutive endocytosis and recycling of uPAR. *PLoS ONE.,* 2008, 3(11), e3730.

Cunningham, O; Andolfo, A; Santovito, ML; Iuzzolino, L; Blasi, F; Sidenius, N. Dimerization controls the lipid rafts partitioning of uPAR/CD87 and regulates its biological functions. *EMBO J,* 2003, Nov, 17, 22(22), 5994-6003.

Czekay, RP; Kuemmel, TA; Orlando, RA; Farquhar, MG. Direct binding of occupied urokinase receptor (uPAR) to LDL receptor-related protein is required for endocytosis of uPAR and regulation of cell surface urokinase activity. *Mol Biol Cell,* 2001, May, 12(5), 1467-1479.

Czekay, RP; Aertgeerts, K; Curriden, SA; Loskutoff, DJ. Plasminogen activator inhibitor-1 detaches cells from extracellular matrices by inactivating integrins. *J Cell Biol,* 2003, Mar 3, 160(5), 781-791.

Danilkovitch-Miagkova, A; Angeloni, D; Skeel, A; Donley, S; Lerman, M; Leonard, EJ. Integrin-mediated RON growth factor receptor phosphorylation requires tyrosine kinase activity of both the receptor and c-Src. *J Biol Chem,* 2000, May 19, 275(20), 14783-14786.

Davis, EM; Trinkaus, JP. Significance of cell-to cell contacts for the directional movement of neural crest cells within a hydrated collagen lattice. *J Embryol Exp Morphol,* 1981, Jun;63:29-51.

de Paulis, A; Montuori, N; Prevete, N; Fiorentino, I; Rossi, FW; Visconte, V; Rossi, G; Marone, G; Ragno, P. Urokinase induces basophil chemotaxis through a urokinase receptor epitope that is an endogenous ligand for formyl peptide receptor-like 1 and -like 2. *J Immunol,* 2004, Nov 1, 173(9), 5739-5748.

Degryse, B. Is uPAR the centre of a sensing system involved in the regulation of inflammation? *Curr Med Chem-Anti-Inflammatory & Anti-Allergy Agents,* 2003, 2(3), 237-259.

Degryse, B. The urokinase receptor (uPAR) and integrins constitute a cell migration signalosome. In: Dylan Edwards, Gunilla Hoyer-Hansen, Francesco Blasi, Bonnie F. Sloane editors. *The Cancer Degradome: Proteases in Cancer Biology*. New-York: Springer; 2008, 451-474.

Degryse, B; Resnati, M; Rabbani, SA; Villa, A; Fazioli, F; Blasi, F. Src-dependence and pertussis-toxin sensitivity of urokinase receptor-dependent chemotaxis and cytoskeleton reorganization in rat smooth muscle cells. *Blood,* 1999, Jul 15, 94(2), 649-662.

Degryse, B; Orlando, S; Resnati, M; Rabbani, SA; Blasi, F. Urokinase/urokinase receptor and vitronectin/$\alpha v\beta 3$ integrin induce chemotaxis and cytoskeleton reorganization through different signalling pathways. *Oncogene,* 2001a, Apr 12, 20(16), 2032-2043.

Degryse, B; Sier, CF; Resnati, M; Conese, M; Blasi, F. PAI-1 inhibits urokinase-induced chemotaxis by internalizing the urokinase receptor. *FEBS Lett,* 2001b, Sep, 14, 505(2), 249-254.

Degryse, B; Bonaldi, T; Scaffidi, P; Müller, S; Resnati, M; Sanvito, F; Arrigoni, G; Bianchi, ME. The high mobility group (HMG) boxes of the nuclear protein HMG1 induce chemotaxis and cytoskeleton reorganization in rat smooth muscle cells. *J Cell Biol,* 2001c, Mar 19, 152(6), 1197-1206.

Degryse, B; Neels, JG; Czekay, RP; Aertgeerts, K; Kamikubo, Y; Loskutoff, DJ. The low density lipoprotein receptor-related protein is a motogenic receptor for plasminogen activator inhibitor-1. *J Biol Chem,* 2004, May, 21, 279(21), 22595-22604.

Degryse, B; Resnati, M; Czekay, RP; Loskutoff, DJ; Blasi, F. Domain 2 of the urokinase receptor contains an integrin-interacting epitope with intrinsic signalling activity: generation of a new integrin inhibitor. *J Biol Chem,* 2005, Jul 1, 280(26), 24792-24803.

Degryse, B; Eden, G; Arnaudova, R; Furlan, F; Blasi, F. Identification of a Mitotic Epitope in the Domain 2 of the Urokinase Receptor (uPAR). In: 11th International Workshop on *Molecular and Cellular Biology of Plasminogen Activation.* Vår Gård Saltsjöbaden, Sweden, 16-20th June, 2007, Abstract 068.

Degryse, B; Fernandez-Recio, J; Citro, V; Blasi, F; Cubellis, MV. In silico docking of urokinase plasminogen activator and integrins. *BMC Bioinformatics,* 2008, Mar 26, 9 Suppl 2:S8.

Demetriou, MC; Cress, AE. Integrin clipping: a novel adhesion switch? *J Cell Biochem,* 2004, Jan 1, 91(1), 26-35.

Demetriou, MC; Pennington, ME; Nagle, RB; Cress, AE. Extracellular alpha 6 integrin cleavage by urokinase-type plasminogen activator in human prostate cancer. *Exp Cell Res,* 2004, Apr 1, 294(2), 550-558.

Dumler, I; Weis, A; Mayboroda, OA; Maasch, C; Jerke, U; Haller, H; Gulba, DC. The Jak/Stat pathway and urokinase receptor signalling in human aortic vascular smooth muscle cells. *J Biol Chem,* 1998, Jan 2, 273(1), 315-321.

Dumler, I; Kopmann, A; Weis, A; Mayboroda, OA; Wagner, K; Gulba, DC; Haller, H. Urokinase activates the Jak/Stat signal transduction pathway in human vascular endothelial cells. *Arterioscler Thromb Vasc Biol,* 1999a, Feb, 19(2), 290-297.

Dumler, I; Kopmann, A; Wagner, K; Mayboroda, OA; Jerke, U; Dietz, R; Haller, H; Gulba, DC. Urokinase induces activation and formation of Stat4 and Stat1-Stat2 complexes in human vascular smooth muscle cells. *J Biol Chem,* 1999b, Aug 20, 274(34), 24059-24065.

Ellis, V; Scully, MF; Kakkar, VV. Plasminogen activation initiated by single-chain urokinase-type plasminogen activator. Potentiation by U937 monocytes. *J Biol Chem,* 1989, Feb 5, 264(4), 2185-2188.

Ezratty, EJ; Partridge, MA; Gundersen, GG. Microtubule-induced focal adhesion disassembly is mediated by dynamin and focal adhesion kinase. *Nat Cell Biol,* 2005, Jun, 7(6), 581-590.

Fazioli, F; Resnati, M; Sidenius, N; Higashimoto, Y; Appella, E; Blasi, F.A urokinase-sensitive region of the human urokinase receptor is responsible for its chemotactic activity. *EMBO J,* 1997, Dec 15, 16(24), 7279-7286.

Felts, Recherches Expérimentales sur le passage des Leucocytes à travers les parois Vasculaires. *Robins' Journal de l'Anat et de la Physiol,* 1870, January.

Fernandez-Borja, M; van Buul, JD; Hordijk, PL. The regulation of leukocyte transendothelial migration by endothelial signaling events. *Cardiovasc Res,* 2010, In the press.

Ffrench-Constant, C; Colognato, H. Integrins: versatile integrators of extracellular signals. *Trends Cell Biol,* 2004, Dec, 14(12), 678-686.

Franco, SJ; Rodgers, MA; Perrin, BJ; Han, J; Bennin, DA; Critchley, DR; Huttenlocher, A. Calpain-mediated proteolysis of talin regulates adhesion dynamics. *Nat Cell Biol,* 2004, Oct;6(10), 977-983.

Franco, P; Vocca, I; Carriero, MV; Alfano, D; Cito, L; Longanesi-Cattani, I; Grieco, P; Ossowski, L; Stoppelli, MP. Activation of urokinase receptor by a novel interaction between the connecting peptide region of urokinase and alpha v beta 5 integrin. *J Cell Sci,* 2006, Aug 15, 119(Pt 16), 3424-3434.

Friedl, P; Borgmann, S; Bröcker, EB. Amoeboid leukocyte crawling through extracellular matrix: lessons from the Dictyostelium paradigm of cell movement. *J Leukoc Biol,* 2001, Oct;70(4), 491-509.

Friedl, P; Wolf, K. Proteolytic and non-proteolytic migration of tumour cells and leucocytes. *Biochem Soc Symp,* 2003, (70), 277-285.

Friedl, P; Wolf, K. Plasticity of cell migration: a multiscale tuning model. *J Cell Biol,* 2010, Jan 11, 188(1), 11-19.

Friedrichson, T; Kurzchalia, TV. Microdomains of GPI-anchored proteins in living cells revealed by crosslinking. *Nature,* 1998, Aug 20, 394(6695), 802-805.

Furlan, F; Orlando, S; Laudanna, C; Resnati, M; Basso, V; Blasi, F; Mondino, A. The soluble D2D3(88-274) fragment of the urokinase receptor inhibits monocyte chemotaxis and integrin-dependent cell adhesion. *J Cell Sci,* 2004, Jun 15, 117(Pt 14), 2909-2916.

Galbraith, CG; Yamada, KM; Galbraith, JA. Polymerizing actin fibers position integrins primed to probe for adhesion sites. *Science,* 2007, Feb, 16, 315(5814), 992-995.

Gargiulo, L; Longanesi-Cattani, I; Bifulco, K; Franco, P; Raiola, R; Campiglia, P; Grieco, P; Peluso, G; Stoppelli, MP; Carriero, MV. Cross-talk between fMLP and vitronectin receptors triggered by urokinase receptor-derived SRSRY peptide. *J Biol Chem,* 2005, Jul 1, 280(26), 25225-32.

Gellert, GC; Goldfarb, RH; Kitson, RP. Physical association of uPAR with the alphaV integrin on the surface of human NK cells. *Biochem Biophys Res Commun,* 2004, Mar, 19, 315(4), 1025-1032.

Gerthoffer, WT. Migration of airway smooth muscle cells. *Proc Am Thorac Soc,* 2008, Jan 1, 5(1), 97-105.

Ginsberg, MH; Partridge, A; Shattil, SJ. Integrin regulation. *Curr Opin Cell Biol,* 2005, Oct, 17(5), 509-516.

Ghosh, S; Brown, R; Jones, JC; Ellerbroek, SM; Stack, MS. Urinary-type plasminogen activator (uPA) expression and uPA receptor localization are regulated by alpha 3beta 1 integrin in oral keratinocytes. *J Biol Chem,* 2000, Aug 4, 275(31), 23869-23876.

Ghosh, S; Johnson, JJ; Sen, R; Mukhopadhyay, S; Liu, Y; Zhang, F; Wei, Y; Chapman, HA; Stack, MS. Functional relevance of urinary-type plasminogen activator receptor-alpha3beta1 integrin association in proteinase regulatory pathways. *J Biol Chem,* 2006, May, 12, 281(19), 13021-13029.

Gieni, RS; Hendzel, MJ. Mechanotransduction from the ECM to the genome: are the pieces now in place? *J Cell Biochem,* 2008, Aug 15, 104(6), 1964-1987.

Gondi, CS; Kandhukuri, N; Kondraganti, S; Gujrati, M; Olivero, WC; Dinh, DH; Rao, JS. Down-regulation of uPAR and cathepsin B retards cofilin dephosphorylation. *Int J Oncol,* 2006, Mar, 28(3), 633-639.

Gormley, IP; Ross, A. The morphology of phytohaemagglutinin (PHA)-stimulated lymphocytes and permanent lymphoid cell lines seen by the scanning electron microscope. *Eur J Cancer,* 1972, Oct, 8(5), 491-494.

Hapke, S; Gawaz, M; Dehne, K; Köhler, J; Marshall, JF; Graeff, H; Schmitt, M; Reuning, U; Lengyel, E. Beta(3)A-integrin downregulates the urokinase-type plasminogen activator receptor (u-PAR) through a PEA3/ets transcriptional silencing element in the u-PAR promoter. *Mol Cell Biol,* 2001a, Mar; 21(6), 2118-21132.

Hapke, S; Kessler, H; Arroyo de Prada, N; Benge, A; Schmitt, M; Lengyel, E; Reuning, U. Integrin alpha(v)beta(3)/vitronectin interaction affects expression of the urokinase system in human ovarian cancer cells. *J Biol Chem,* 2001b, Jul 13, 276(28), 26340-26348.

Hayem, 1870. Le Mécanisme de la Suppuration. *Archives Gen de Médecine* 1870 March;364-369.

Hooke, R. Micrographia : or some physiological descriptions of minute bodies made by magnifying glasses with observations and inquiries thereupon. London: Martyn, J. and Allestry, *J. printers to the Royal Society*, 1665.

Horwitz, R; Webb, D. Cell migration. *Curr Biol,* 2003, Sep 30, 13(19), R756-R759.

Høyer-Hansen, G; Rønne, E; Solberg, H; Behrendt, N; Ploug, M; Lund, LR; Ellis, V; Danø, K. Urokinase plasminogen activator cleaves its cell surface receptor releasing the ligand-binding domain. *J Biol Chem,* 1992, Sep 5, 267(25), 18224-18229.

Høyer-Hansen, G; Pessara, U; Holm, A; Pass, J; Weidle, U; Danø, K; Behrendt, N. Urokinase-catalysed cleavage of the urokinase receptor requires an intact glycolipid anchor. *Biochem J,* 2001, Sep 15, 358(Pt 3), 673-679.

Huai, Q; Mazar, AP; Kuo, A; Parry, GC; Shaw, DE; Callahan, J; Li, Y; Yuan, C; Bian, C; Chen, L; Furie, B; Furie, BC; Cines, DB; Huang, M. Structure of human urokinase plasminogen activator in complex with its receptor. *Science,* 2006, Feb 3, 311(5761), 656-659.

Huang, M; Mazar, AP; Parry, G; Higazi, AA; Kuo, A; Cines, DB. Crystallization of soluble urokinase receptor (suPAR) in complex with urokinase amino-terminal fragment (1-143). *Acta Crystallogr D Biol Crystallogr,* 2005, Jun, 61(Pt 6), 697-700.

Ilina, O; Friedl, P. Mechanisms of collective cell migration at a glance. *J Cell Sci* 2009, Sep 15, 122(Pt 18), 3203-3208.

Jiang, WG. Membrane ruffling of cancer cells: a parameter of tumour cell motility and invasion. *Eur J Surg Oncol* 1995 Jun; 21(3), 307-309.

Kamikubo, Y; Neels, JG; Degryse, B. Vitronectin inhibits plasminogen activator inhibitor-1-induced signalling and chemotaxis by blocking plasminogen activator inhibitor-1 binding to the low-density lipoprotein receptor-related protein. *Int J Biochem Cell Biol,* 2009, Mar; 41(3), 578-585.

Katoh, K; Kano, Y; Ookawara, S. Role of stress fibers and focal adhesions as a mediator for mechano-signal transduction in endothelial cells in situ. *Vasc Health Risk Manag,* 2008, 4(6), 1273-1282.

Katsumi, A; Orr, AW; Tzima, E; Schwartz, MA. Integrins in mechanotransduction. *J Biol Chem,* 2004, Mar 26, 279(13), 12001-12004.

Katsumi, A; Naoe, T; Matsushita, T; Kaibuchi, K; Schwartz, MA. Integrin activation and matrix binding mediate cellular responses to mechanical stretch. *J Biol Chem,* 2005, Apr 29, 280(17), 16546-16549.

Khatib, AM; Nip, J; Fallavollita, L; Lehmann, M; Jensen, G; Brodt, P. Regulation of urokinase plasminogen activator/plasmin-mediated invasion of melanoma cells by the integrin vitronectin receptor alphaVbeta3. *Int J Cancer,* 2001, Feb 1, 91(3), 300-308.

Kinashi, T. Intracellular signalling controlling integrin activation in lymphocytes. *Nat Rev Immunol,* 2005, Jul, 5(7), 546-559.

Kim, KS; Hong, YK; Joe, YA; Lee, Y; Shin, JY; Park, HE; Lee, IH; Lee, SY; Kang, DK; Chang, SI; Chung, SI. Anti-angiogenic activity of the recombinant kringle domain of urokinase and its specific entry into endothelial cells. *J Biol Chem,* 2003a, Mar 28, 278(13), 11449-11456.

Kim, KS; Hong, YK; Lee, Y; Shin, JY; Chang, SI; Chung, SI; Joe, YA. Differential inhibition of endothelial cell proliferation and migration by urokinase subdomains: amino-terminal fragment and kringle domain. *Exp Mol Med,* 2003, Dec 31, 35(6), 578-585.

Kim, CK; Hong, SH; Joe, YA; Shim, BS; Lee, SK; Hong, YK. The recombinant kringle domain of urokinase plasminogen activator inhibits in vivo malignant glioma growth. *Cancer Sci,* 2007, Feb, 98(2), 253-258.

Kindzelskii, AL; Xue, W; Todd, RF 3rd; Boxer, LA; Petty, HR. Aberrant capping of membrane proteins on neutrophils from patients with leukocyte adhesion deficiency. *Blood,* 1994, Mar 15, 83(6), 1650-1655.

Kindzelskii, AL; Eszes, MM; Todd, RF 3rd; Petty, HR. Proximity oscillations of complement type 4 (alphaX beta2) and urokinase receptors on migrating neutrophils. *Biophys J,* 1997, Oct, 73(4), 1777-1784.

Kjøller, L; Hall, A. Rac mediates cytoskeletal rearrangements and increased cell motility induced by urokinase-type plasminogen activator receptor binding to vitronectin. *J Cell Biol,* 2001, Mar 19, 152(6), 1145-1157.

Koshelnick, Y; Ehart, M; Hufnagl, P; Heinrich, PC; Binder, BR. Urokinase receptor is associated with the components of the JAK1/STAT1 signalling pathway and leads to activation of this pathway upon receptor clustering in the human kidney epithelial tumor cell line TCL-598. *J Biol Chem* 1997, Nov 7, 272(45), 28563-28567.

Kwak, SH; Mitra, S; Bdeir, K; Strassheim, D; Park, JS; Kim, JY; Idell, S; Cines, D; Abraham, E. The kringle domain of urokinase-type plasminogen activator potentiates LPS-induced neutrophil activation through interaction with {alpha}V{beta}3 integrins. *J Leukoc Biol,* 2005, Oct, 78(4), 937-945.

Lämmermann, T; Sixt, M. Mechanical modes of 'amoeboid' cell migration. *Curr Opin Cell Biol,* 2009, Oct, 21(5), 636-644.

Lauffenburger, DA; Horwitz, AF. Cell migration: a physically integrated molecular process. *Cell,* 1996, Feb 9, 84(3), 359-369.

Le Clainche, C; Carlier, MF. Regulation of actin assembly associated with protrusion and adhesion in cell migration. *Physiol Rev,* 2008, Apr;88(2), 489-513.

Leksa, V; Godár, S; Cebecauer, M; Hilgert, I; Breuss, J; Weidle, UH; Horejsí, V; Binder, BR; Stockinger, H. The N terminus of mannose 6-phosphate/insulin-like growth factor 2 receptor in regulation of fibrinolysis and cell migration. *J Biol Chem,* 2002, Oct 25, 277(43), 40575-40582.

Li, H; Kuo, A; Kochan, J; Strickland, D; Kariko, K; Barnathan, ES; Cines, DB. Endocytosis of urokinase-plasminogen activator inhibitor type 1 complexes bound to a chimeric transmembrane urokinase receptor. *J Biol Chem,* 1994, Mar 18, 269(11), 8153-8158.

Li, S; Huang, NF; Hsu, S. Mechanotransduction in endothelial cell migration. *J Cell Biochem,* 2005, Dec 15, 96(6), 1110-1126.

Liso, V; Specchia, G; Pavone, V; Colotta, F; Riezzo, A; Ferrannini, A; Tursi, A. Acute lymphoblastic leukemia hand-mirror cells. Study of nine cases. *Blut,* 1983, Nov, 47(5), 297-306.

Liu, HW; Luo, YC; Ho, CL; Yang, JY; Lin, CH. Locomotion guidance by extracellular matrix is adaptive and can be restored by a transient change in Ca2+ level. *PLoS One,* 2009, Oct 5, 4(10), e7330.

Llinas, P; Le Du, MH; Gårdsvoll, H; Danø, K; Ploug, M; Gilquin, B; Stura, EA; Ménez, A. Crystal structure of the human urokinase plasminogen activator receptor bound to an antagonist peptide. *EMBO J,* 2005, May 4, 24(9), 1655-1663.

Madsen, CD; Ferraris, GM; Andolfo, A; Cunningham, O; Sidenius, N. uPAR-induced cell adhesion and migration: vitronectin provides the key. *J Cell Biol,* 2007, Jun 4, 177(5), 927-939.

Malengo, G; Andolfo, A; Sidenius, N; Gratton, E; Zamai, M; Caiolfa, VR. Fluorescence correlation spectroscopy and photon counting histogram on membrane proteins: functional dynamics of the glycosylphosphati dylinositol-anchored urokinase plasminogen activator receptor. *J Biomed Opt,* 2008, May-Jun;13(3), 031215.

Margheri, F; Manetti, M; Serratì, S; Nosi, D; Pucci, M; Matucci-Cerinic, M; Kahaleh, B; Bazzichi L; Fibbi, G; Ibba-Manneschi, L; Del Rosso, M. Domain 1 of the urokinase-type plasminogen activator receptor is required for its morphologic and functional, beta2 integrin-mediated connection with actin cytoskeleton in human microvascular endothelial cells: failure of association in systemic sclerosis endothelial cells. *Arthritis Rheum,* 2006, Dec, 54(12), 3926-3938.

May, AE; Kanse, SM; Lund, LR; Gisler, RH; Imhof, BA; Preissner, KT. Urokinase receptor (CD87) regulates leukocyte recruitment via beta 2 integrins in vivo. *J Exp Med,* 1998, Sep 21, 188(6), 1029-1037.

May, AE; Neumann, FJ; Schömig, A; Preissner, KT. VLA-4 (alpha(4)beta(1)) engagement defines a novel activation pathway for beta(2) integrin-dependent leukocyte adhesion involving the urokinase receptor. *Blood,* 2000, Jul 15, 96(2), 506-513.

May, AE; Schmidt, R; Kanse, SM; Chavakis, T; Stephens, RW; Schömig, A; Preissner, KT; Neumann, FJ. Urokinase receptor surface expression regulates monocyte adhesion in acute myocardial infarction. *Blood,* 2002, Nov, 15, 100(10), 3611-3617.

Mazar, AP. Urokinase plasminogen activator receptor choreographs multiple ligand interactions: implications for tumor progression and therapy. *Clin Cancer Res,* 2008, Sep 15, 14(18), 5649-5655.

Mazzarello, P. A unifying concept: the history of cell theory. *Nat Cell Biol,* 1999, 1, E13-E15.

Mazzieri, R; D'Alessio, S; Kenmoe, RK; Ossowski, L; Blasi, F. An uncleavable uPAR mutant allows dissection of signaling pathways in uPA-dependent cell migration. *Mol Biol Cell,* 2006, Jan, 17(1), 367-378.

McNiven, MA; Baldassarre, M; Buccione, R. The role of dynamin in the assembly and function of podosomes and invadopodia. *Front Biosci,* 2004, May, 1, 9, 1944-1953.

Meyers, J; Craig, J; Odde, DJ. Potential for control of signaling pathways via cell size and shape. *Curr Biol,* 2006, Sep 5, 16(17), 1685-1693.

Miyamoto, S; Teramoto, H; Gutkind, JS; Yamada, KM. Integrins can collaborate with growth factors for phosphorylation of receptor tyrosine kinases and MAP kinase activation: roles of integrin aggregation and occupancy of receptors. *J Cell Biol,* 1996, Dec, 135(6 Pt 1), 1633-1642.

Montuori, N; Rossi, G; Ragno, P. Cleavage of urokinase receptor regulates its interaction with integrins in thyroid cells. *FEBS Lett,* 1999, Oct 22, 460(1), 32-36.

Montuori, N; Carriero, MV; Salzano, S; Rossi, G; Ragno, P. The cleavage of the urokinase receptor regulates its multiple functions. *J Biol Chem,* 2002, Dec, 6, 277(49), 46932-46939.

Mustjoki, S; Alitalo, R; Stephens, RW; Vaheri, A. Blast cell-surface and plasma soluble urokinase receptor in acute leukemia patients: relationship to classification and response to therapy. *Thromb Haemost,* 1999, May, 81(5), 705-710.

Myöhänen, HT; Stephens, RW; Hedman, K; Tapiovaara, H; Rønne, E; Høyer-Hansen, G; Danø, K; Vaheri, A. Distribution and lateral mobility of the urokinase-receptor complex at the cell surface. *J Histochem Cytochem,* 1993, Sep, 41(9), 1291-1301.

Nadarajah, B; Alifragis, P; Wong, RO; Parnavelas, JG. Neuronal migration in the developing cerebral cortex: observations based on real-time imaging. *Cereb Cortex,* 2003, Jun, 13(6), 607-611.

Naldini, L; Tamagnone, L; Vigna, E; Sachs, M; Hartmann, G; Birchmeier, W; Daikuhara, Y; Tsubouchi, H; Blasi, F; Comoglio, PM. Extracellular proteolytic cleavage by urokinase is required for activation of hepatocyte growth factor/scatter factor. *EMBO J,* 1992, Dec;11(13), 4825-4833.

Naldini, L; Vigna, E; Bardelli, A; Follenzi, A; Galimi, F; Comoglio, PM. Biological activation of pro-HGF (hepatocyte growth factor) by urokinase is controlled by a stoichiometric reaction. *J Biol Chem,* 1995, Jan, 13, 270(2), 603-611.

Nguyen, DH; Catling, AD; Webb, DJ; Sankovic, M; Walker, LA; Somlyo, AV; Weber, MJ; Gonias, SL. Myosin light chain kinase functions downstream of Ras/ERK to promote migration of urokinase-type plasminogen activator-stimulated cells in an integrin-selective manner. *J Cell Biol,* 1999, Jul, 12, 146(1), 149-164.

Nicholson, TB; Stanners, CP. Specific inhibition of GPI-anchored protein function by homing and self-association of specific GPI anchors. *J Cell Biol,* 2006, Nov 20, 175(4), 647-659.

Nip, J; Rabbani, SA; Shibata, HR; Brodt, P. Coordinated expression of the vitronectin receptor and the urokinase-type plasminogen activator receptor in metastatic melanoma cells. *J Clin Invest,* 1995, May;95(5), 2096-103.

Norberg, B; Gippert, H; Norberg, A; Parkhede, U; Rydgren, L. Locomotion and phagocytosis of polymorphonuclear leukocytes (PMNs) in infected urine. Reversal of PMN polarity. *Eur Urol,* 1978, 4(6), 438-440.

Nykjaer, A; Petersen, CM; Møller, B; Jensen, PH; Moestrup, SK; Holtet, TL; Etzerodt, M; Thøgersen, HC; Munch, M; Andreasen, PA; Gliemann, J. Purified alpha 2-macroglobulin receptor/LDL receptor-related protein binds urokinase.plasminogen activator inhibitor type-1 complex. Evidence that the alpha 2-macroglobulin receptor mediates cellular degradation of urokinase receptor-bound complexes. *J Biol Chem,* 1992, Jul 25, 267(21), 14543-14546.

Nykjaer, A; Kjøller, L; Cohen, RL; Lawrence, DA; Garni-Wagner, BA; Todd, RF 3rd; van Zonneveld, AJ; Gliemann, J; Andreasen, PA. Regions involved in binding of urokinase-type-1 inhibitor complex and pro-urokinase to the endocytic alpha 2-macroglobulin receptor/low density lipoprotein receptor-related protein. Evidence that the urokinase receptor protects pro-urokinase against binding to the endocytic receptor. *J Biol Chem,* 1994, Oct, 14, 269(41), 25668-25676.

Nykjaer, A; Conese, M; Christensen, EI; Olson, D; Cremona, O; Gliemann, J; Blasi, F. Recycling of the urokinase receptor upon internalization of the uPA:serpin complexes. *EMBO J,* 1997, May 15, 16(10), 2610-2620.

Nykjaer, A; Christensen, EI; Vorum, H; Hager, H; Petersen, CM; Røigaard, H; Min, HY; Vilhardt, F; Møller, LB; Kornfeld, S; Gliemann, J. Mannose 6-phosphate/insulin-like growth factor-II receptor targets the urokinase receptor to lysosomes via a novel binding interaction. *J Cell Biol,* 1998, May 4, 141(3), 815-828.

Odekon, LE; Sato, Y; Rifkin, DB. Urokinase-type plasminogen activator mediates basic fibroblast growth factor-induced bovine endothelial cell migration independent of its proteolytic activity. *J Cell Physiol,* 1992, Feb;150(2), 258-263.

Odekon, LE; Blasi, F; Rifkin, DB. Requirement for receptor-bound urokinase in plasmin-dependent cellular conversion of latent TGF-beta to TGF-beta. *J Cell Physiol,* 1994, Mar;158(3), 398-407.

Okada, SS; Tomaszewski, JE; Barnathan, ES. Migrating vascular smooth muscle cells polarize cell surface urokinase receptors after injury in vitro. *Exp Cell Res,* 1995, Mar, 217(1), 180-187.

Okumura, Y; Kamikubo, Y; Curriden, SA; Wang, J; Kiwada, T; Futaki, S; Kitagawa, K; Loskutoff, DJ. Kinetic analysis of the interaction between vitronectin and the urokinase receptor. *J Biol Chem,* 2002, Mar 15, 277(11), 9395-9404.

Olson, MF; Sahai E. The actin cytoskeleton in cancer cell motility. *Clin Exp Metastasis,* 2009, 26(4), 273-287.

Orlean, P; Menon AK. Thematic review series: lipid posttranslational modifications. GPI anchoring of protein in yeast and mammalian cells, or: how we learned to stop worrying and love glycophospholipids. *J Lipid Res,* 2007, May;48(5), 993-1011.

Paulick, MG; Bertozzi CR. The glycosylphosphatidylinositol anchor: a complex membrane-anchoring structure for proteins. *Biochemistry,* 2008, Jul 8, 47(27), 6991-7000.

Pawar, SC; Demetriou, MC; Nagle, RB; Bowden, GT; Cress, AE. Integrin alpha6 cleavage: a novel modification to modulate cell migration. *Exp Cell Res,* 2007, Apr 1, 313(6), 1080-1089.

Petty, HR; Kindzelskii, AL; Adachi, Y; Todd, RF 3rd. Ectodomain interactions of leukocyte integrins and pro-inflammatory GPI-linked membrane proteins. *J Pharm Biomed Anal,* 1997, Jun, 15(9-10), 1405-1416.

Ploug, M; Ellis, V. Structure-function relationships in the receptor for urokinase-type plasminogen activator. Comparison to other members of the Ly-6 family and snake venom alpha-neurotoxins. *FEBS Lett,* 1994, Aug 1, 349(2), 163-168.

Pluskota, E; Soloviev, DA; Plow, EF. Convergence of the adhesive and fibrinolytic systems: recognition of urokinase by integrin alpha Mbeta 2 as well as by the urokinase receptor regulates cell adhesion and migration. *Blood,* 2003, Feb, 15, 101(4), 1582-1590.

Pluskota, E; Soloviev, DA; Bdeir, K; Cines, DB; Plow EF. Integrin alphaMbeta2 orchestrates and accelerates plasminogen activation and fibrinolysis by neutrophils. *J Biol Chem,* 2004, Apr 23, 279(17), 18063-18072.

Pöllänen, J; Hedman, K; Nielsen, LS; Danø, K; Vaheri, A. Ultrastructural localization of plasma membrane-associated urokinase-type plasminogen activator at focal contacts. *J Cell Biol,* 1988, Jan;106(1), 87-95.

Porter, JC; Hogg, N. Integrins take partners: cross-talk between integrins and other membrane receptors. *Trends Cell Biol.*, 1998, Oct, 8(10), 390-396.

Prager, GW; Breuss, JM; Steurer, S; Olcaydu, D; Mihaly, J; Brunner, PM; Stockinger, H; Binder, BR. Vascular endothelial growth factor receptor-2-induced initial endothelial cell migration depends on the presence of the urokinase receptor. *Circ Res,* 2004, Jun 25, 94(12), 1562-1570.

Ragno, P. The urokinase receptor: a ligand or a receptor? Story of a sociable molecule. *Cell Mol Life Sci,* 2006, May; 63(9), 1028-1037.

Ragno, P; Montuori, N; Covelli, B; Hoyer-Hansen, G; Rossi, G. Differential expression of a truncated form of the urokinase-type plasminogen-activator receptor in normal and tumor thyroid cells. *Cancer Res,* 1998, Mar 15, 58(6), 1315-1319.

Reinartz, J; Schäfer, B; Batrla, R; Klein, CE; Kramer, MD. Plasmin abrogates alpha v beta 5-mediated adhesion of a human keratinocyte cell line (HaCaT) to vitronectin. *Exp Cell Res,* 1995, Oct; 220(2), 274-282.

Resnati, M; Guttinger, M; Valcamonica, S; Sidenius, N; Blasi, F; Fazioli, F. Proteolytic cleavage of the urokinase receptor substitutes for the agonist-induced chemotactic effect. *EMBO J,* 1996, Apr 1, 15(7), 1572-1582.

Resnati, M; Pallavicini, I; Wang, JM; Oppenheim, J; Serhan, CN; Romano, M; Blasi, F. The fibrinolytic receptor for urokinase activates the G protein-coupled chemotactic receptor FPRL1/LXA4R. *Proc Natl Acad Sci USA,* 2002, Feb 5, 99(3), 1359-1364.

Ridley, AJ; Schwartz, MA; Burridge, K; Firtel, RA; Ginsberg, MH; Borisy, G; Parsons, JT; Horwitz, AR. Cell migration: integrating signals from front to back. *Science,* 2003, Dec 5, 302(5651), 1704-1709.

Roca-Cusachs, P; Gauthier, NC; Del Rio, A; Sheetz, MP. Clustering of alpha(5)beta(1) integrins determines adhesion strength whereas alpha(v)beta(3) and talin enable mechanotransduction. *Proc Natl Acad Sci* USA, 2009, Sep 22, 106(38), 16245-16250.

Rørth, P. Collective cell migration. *Annu Rev Cell Dev Biol,* 2009, 25, 407-429.

Sadowski, L; Pilecka, I; Miaczynska, M. Signaling from endosomes: location makes a difference. *Exp Cell Res,* 2009, May 15, 315(9), 1601-1609.

Sage, PT; Carman CV. Settings and mechanisms for trans-cellular diapedesis. *Front Biosci.,* 2009, Jun 1, 14, 5066-5083.

Sahores, M; Prinetti, A; Chiabrando, G; Blasi F; Sonnino, S. uPA binding increases UPAR localization to lipid rafts and modifies the receptor microdomain composition. *Biochim Biophys Acta,* 2008, Jan, 1778(1), 250-259.

Salasznyk, RM; Zappala, M; Zheng, M; Yu, L; Wilkins-Port, C; McKeown-Longo, PJ. The uPA receptor and the somatomedin B region of vitronectin direct the localization of uPA to focal adhesions in microvessel endothelial cells. *Matrix Biol,* 2007, Jun;26(5), 359-370.

Sawai, H; Okada, Y; Funahashi, H; Matsuo, Y; Takahashi, H; Takeyama, H; Manabe, T. Interleukin-1alpha enhances the aggressive behavior of pancreatic cancer cells by regulating the alpha6beta1-integrin and urokinase plasminogen activator receptor expression. *BMC Cell Biol,* 2006, Feb 20, 7, 8.

Santoro, MM; Gaudino, G. Cellular and molecular facets of keratinocyte reepithelization during wound healing. *Exp Cell Res,* 2005, Mar 10, 304(1), 274-286.

Schleiden, MJ. *Arch Anat Physiol Wiss Med,* 1838, 13, 137-176.

Schmitt-Gräff, A; Fischer, JT. Hand-mirror forms in AUL: cytochemistry and ultrastructure. *Blut,* 1983, Apr;46(4), 227-229.

Schober, A; Weber, C. Mechanisms of monocyte recruitment in vascular repair after injury. *Antioxid Redox Signal,* 2005, Sep-Oct, 7(9-10), 1249-1257.

Schreiner, GF; Unanue, ER. The modulation of spontaneous and anti-Ig-stimulated motility of lymphocytes by cyclic nucleotides and adrenergic and cholinergic agents. *J Immunol,* 1975, Feb;114(2 pt 2), 802-808.

Schvartz, I; Seger, D; Shaltiel, S. Vitronectin *Int J Biochem Cell Biol,* 1999, May, 31(5), 539-544.

Schwab, W; Gavlik, JM; Beichler, T; Funk, RH; Albrecht, S; Magdolen, V; Luther, T; Kasper, M; Shakibaei, M. Expression of the urokinase-type plasminogen activator receptor in human articular chondrocytes: association with caveolin and beta 1-integrin. *Histochem Cell Biol,* 2001, Apr;115(4), 317-323.

Schwann, T. *Mikroskopische Untersuchungen über die Übereinstimmung in der Struktur und dem Wachstum der Tiere und Pflanzen.* Sander'schen Buchhandlung, Berlin, 1839.

Seiffert, D; Loskutoff, DJ. Evidence that type 1 plasminogen activator inhibitor binds to the somatomedin B domain of vitronectin. *J Biol Chem,* 1991, Feb 15, 266(5), 2824-2830.

Selleri, C; Montuori, N; Ricci, P; Visconte, V; Carriero, MV; Sidenius, N; Serio, B; Blasi, F; Rotoli, B; Rossi, G; Ragno, P. Involvement of the urokinase-type plasminogen activator receptor in hematopoietic stem cell mobilization. *Blood,* 2005, Mar 1, 105(5), 2198-2205.

Sid, B; Dedieu, S; Delorme, N; Sartelet, H; Rath, GM; Bellon, G; Martiny, L. Human thyroid carcinoma cell invasion is controlled by the low density lipoprotein receptor-related protein-mediated clearance of urokinase plasminogen activator. *Int J Biochem Cell Biol,* 2006, 38(10), 1729-1740.

Sidenius, N; Blasi, F. Domain 1 of the urokinase receptor (uPAR) is required for uPAR-mediated cell binding to vitronectin. *FEBS Lett,* 2000, Mar 17, 470(1), 40-46.

Sidenius, N; Andolfo, A; Fesce, R; Blasi, F. Urokinase regulates vitronectin binding by controlling urokinase receptor oligomerization. *J Biol Chem,* 2002, Aug 2, 277(31), 27982-27990.

Sier, CF; Sidenius, N; Mariani, A; Aletti, G; Agape, V; Ferrari, A; Casetta, G; Stephens, RW; Brünner, N; Blasi F.Presence of urokinase-type plasminogen activator receptor in urine of cancer patients and its possible clinical relevance. *Lab Invest,* 1999, Jun, 79(6), 717-722.

Silver, FH; Siperko, LM. Mechanosensing and mechanochemical transduction: how is mechanical energy sensed and converted into chemical energy in an extracellular matrix? *Crit Rev Biomed Eng,* 2003, 31(4), 255-331.

Silvestri, I; Longanesi Cattani, I; Franco, P; Pirozzi, G; Botti, G; Stoppelli, MP; Carriero, MV. Engaged urokinase receptors enhance tumor breast cell migration and invasion by upregulating alpha(v)beta5 vitronectin receptor cell surface expression. *Int J Cancer,* 2002, Dec 20, 102(6), 562-571.

Simon, DI; Rao, NK; Xu, H; Wei, Y; Majdic, O; Ronne, E; Kobzik, L; Chapman, HA. Mac-1 (CD11b/CD18) and the urokinase receptor (CD87) form a functional unit on monocytic cells. *Blood,* 1996, Oct 15, 88(8), 3185-3194.

Simon, DI; Wei, Y; Zhang, L; Rao, NK; Xu, H; Chen, Z; Liu, Q; Rosenberg, S; Chapman, HA. Identification of a urokinase receptor-integrin interaction site. Promiscuous regulator of integrin function. *J Biol Chem,* 2000, Apr 7;275(14), 10228-10234.

Simons, K; Toomre, D. Lipid rafts and signal transduction. *Nat Rev Mol Cell Biol,* 2000, Oct;1(1), 31-39.

Sitrin, RG; Todd, RF 3rd; Albrecht, E; Gyetko, MR. The urokinase receptor (CD87) facilitates CD11b/CD18-mediated adhesion of human monocytes. *J Clin Invest,* 1996, Apr, 15, 97(8), 1942-1951.

Sitrin, RG; Pan, PM; Harper, HA; Todd, RF 3rd; Harsh, DM; Blackwood, RA. Clustering of urokinase receptors (uPAR; CD87) induces proinflammatory signalling in human polymorphonuclear neutrophils. *J Immunol,* 2000, Sep, 15, 165(6), 3341-3349.

Sitrin, RG; Johnson, DR; Pan, PM; Harsh, DM; Huang, J; Petty, HR; Blackwood, RA. Lipid rafts compartmentalization of urokinase receptor signalling in human neutrophils. *Am J Respir Cell Mol Biol,* 2004, Feb, 30(2), 233-241.

Smirnova, T; Segall, JE. Amoeboid chemotaxis: future challenges and opportunities. *Cell Adh Migr,* 2007, Oct;1(4), 165-170.

Stahl, A; Mueller, BM. The urokinase-type plasminogen activator receptor, a GPI-linked protein, is localized in caveolae. *J Cell Biol,* 1995, Apr;129(2), 335-344.

Stahl, A; Mueller, BM. Melanoma cell migration on vitronectin: regulation by components of the plasminogen activation system. *Int J Cancer,* 1997, Mar 28, 71(1), 116-122.

Stepanova, V; Jerke, U; Sagach, V; Lindschau, C; Dietz, R; Haller, H; Dumler, I. Urokinase-dependent human vascular smooth muscle cell adhesion requires selective vitronectin phosphorylation by ectoprotein kinase CK2. *J Biol Chem,* 2002, Mar 22, 277(12), 10265-10272.

Stephens, RW; Nielsen, HJ; Christensen, IJ; Thorlacius-Ussing, O; Sørensen, S; Danø, K; Brünner, N.Plasma urokinase receptor levels in patients with colorectal cancer: relationship to prognosis. *J Natl Cancer Inst,* 1999, May, 19, 91(10), 869-874.

Streuli, CH; Akhtar, N. Signal co-operation between integrins and other receptor systems. *Biochem J,* 2009, Mar 15, 418(3), 491-506.

Szczepanowska, J. Involvement of Rac/Cdc42/PAK pathway in cytoskeletal rearrangements. *Acta Biochim Pol,* 2009, 56(2), 225-234.

Tang, CH; Hill, ML; Brumwell, AN; Chapman, HA; Wei, Y. Signaling through urokinase and urokinase receptor in lung cancer cells requires interactions with beta1 integrins. *J Cell Sci,* 2008, Nov, 15, 121(Pt 22), 3747-3756.

Tang, ML; Vararattanavech, A; Tan, SM. Urokinase-type plasminogen activator receptor induces conformational changes in the integrin alphaMbeta2 headpiece and reorientation of its transmembrane domains. *J Biol Chem,* 2008, Sep 12, 283(37), 25392-25403.

Tarui, T; Mazar, AP; Cines, DB; Takada, Y. Urokinase-type plasminogen activator receptor (CD87) is a ligand for integrins and mediates cell-cell interaction. *J Biol Chem,* 2001a, Feb 9, 276(6), 3983-3990.

Tarui, T; Miles, LA ; Takada, Y. Specific interaction of angiostatin with integrin alpha(v)beta(3) in endothelial cells. *J Biol Chem,* 2001b, Oct, 26, 276(43), 39562-39568.

Tarui, T; Akakura, N; Majumdar, M; Andronicos, N; Takagi, J; Mazar, AP; Bdeir, K; Kuo, A; Yarovoi, SV; Cines, DB; Takada, Y. Direct interaction of the kringle domain of urokinase-type plasminogen activator (uPA) and integrin alpha v beta 3 induces signal transduction and enhances plasminogen activation. *Thromb Haemost,* 2006, Mar, 95(3), 524-534.

Teddy, JM; Kulesa, PM. In vivo evidence for short- and long-range cell communication in cranial neural crest cells. *Development,* 2004, Dec, 131(24), 6141-6151.

Thomas, WJ; Duval-Arnould, B; Creegan, WJ; Schumacher, HR; Forman, DS; Strong DM. Morphologic observations of contact-induced lysis of EBV-infected B lymphocytes by autologous hand mirror T cells. *Am J Hematol,* 1982, Apr, 12(2), 109-119.

Van Haastert, PJ; Veltman, DM. Chemotaxis: navigating by multiple signaling pathways. *Sci STKE,* 2007, Jul 24, 2007(396), pe40.

Varma, R; Mayor, S. GPI-anchored proteins are organized in submicron domains at the cell surface. *Nature* 1998 Aug 20;394(6695), 798-801.

Vicente-Manzanares, M; Webb, DJ; Horwitz, AR. Cell migration at a glance. *J Cell Sci,* 2005, Nov 1, 118(Pt 21), 4917-4919.

Waller, Microscopic Observations on the Perforation of the Capillaries by the Blood, and on the Origin of Mucous and Pus Globules. *Philosophical Magazine,* 1846.

Waltz, DA; Chapman, HA. Reversible cellular adhesion to vitronectin linked to urokinase receptor occupancy. *J Biol Chem,* 1994, May 20, 269(20), 14746-14750.

Wang, GJ; Collinge, M; Blasi, F; Pardi, R; Bender, JR. Posttranscriptional regulation of urokinase plasminogen activator receptor messenger RNA levels by leukocyte integrin engagement. *Proc Natl Acad Sci* U S A, 1998, May 26, 95(11), 6296-6301.

Wang, N; Planus, E; Pouchelet, M; Fredberg, JJ; Barlovatz-Meimon, G. Urokinase receptor mediates mechanical force transfer across the cell surface. *Am J Physiol,* 1995, Apr, 268(4Pt1), C1062-C1066.

Wang, N; Tytell, JD; Ingber, DE. Mechanotransduction at a distance: mechanically coupling the extracellular matrix with the nucleus. *Nat Rev Mol Cell Biol,* 2009, Jan, 10(1), 75-82.

Weaver, AM; Hussaini, IM; Mazar, A; Henkin, J; Gonias, SL. Embryonic fibroblasts that are genetically deficient in low density lipoprotein receptor-related protein demonstrate increased activity of the urokinase receptor system and accelerated migration on vitronectin. *J Biol Chem,* 1997, May 30, 272(22), 14372-14379.

Webb, DJ; Nguyen, DH; Sankovic, M; Gonias, SL. The very low density lipoprotein receptor regulates urokinase receptor catabolism and breast cancer cell motility in vitro. *J Biol Chem,* 1999, Mar 12, 274(11), 7412-7420.

Webb, DJ; Nguyen, DH; Gonias, SL. Extracellular signal-regulated kinase functions in the urokinase receptor-dependent pathway by which neutralization of low density lipoprotein receptor-related protein promotes fibrosarcoma cell migration and matrigel invasion. *J Cell Sci,* 2000, Jan, 113(Pt1), 123-134.

Wei, Y; Waltz, DA; Rao, N; Drummond, RJ; Rosenberg, S; Chapman, HA. Identification of the urokinase receptor as an adhesion receptor for vitronectin. *J Biol Chem,* 1994, Dec, 23, 269(51), 32380-32388.

Wei, Y; Lukashev, M; Simon, DI; Bodary, SC; Rosenberg, S; Doyle, MV; Chapman, HA. Regulation of integrin function by the urokinase receptor. *Science,* 1996, Sep 13, 273(5281), 1551-1555.

Wei, Y; Yang X; Liu, Q; Wilkins, JA; Chapman, HA. A role for caveolin and the urokinase receptor in integrin-mediated adhesion and signalling. *J Cell Biol,* 1999, Mar, 22, 144(6), 1285-1294.

Wei, Y; Eble, JA; Wang, Z; Kreidberg, JA; Chapman, HA. Urokinase receptors promote beta1 integrin function through interactions with integrin alpha3beta1. *Mol Biol Cell,* 2001, Oct;12(10), 2975-2986.

Wei, Y; Czekay, RP; Robillard, L; Kugler, MC; Zhang, F; Kim, KK; Xiong, JP; Humphries, MJ; Chapman, HA. Regulation of alpha5beta1 integrin conformation and function by urokinase receptor binding. *J Cell Biol,* 2005, Jan 31, 168(3), 501-511.

Wei, Y; Tang, CH; Kim, Y; Robillard, L; Zhang, F; Kugler, MC; Chapman, HA. Urokinase receptors are required for alpha 5 beta 1 integrin-mediated signalling in tumor cells. *J Biol Chem,* 2007, Feb 9, 282(6), 3929-3939.

Weijer, CJ. Collective cell migration in development. *J Cell Sci,* 2009, Sep 15, 122(Pt 18), 3215-3223.

Wolf, K; Mazo, I; Leung, H; Engelke, K; von Andrian, UH; Deryugina, EI; Strongin, AY; Bröcker, EB; Friedl, P. Compensation mechanism in tumor cell migration: mesenchymal-amoeboid transition after blocking of pericellular proteolysis. *J Cell Biol,* 2003, Jan 20, 160(2), 267-277.

Wong, WS; Simon, DI; Rosoff, PM; Rao, NK; Chapman, HA. Mechanisms of pertussis toxin-induced myelomonocytic cell adhesion: role of Mac-1(CD11b/CD18) and urokinase receptor (CD87). *Immunology,* 1996, May;88(1), 90-97.

Xia, Y; Borland, G; Huang, J; Mizukami, IF; Petty, HR; Todd, RF 3rd; Ross, GD. Function of the lectin domain of Mac-1/complement receptor type 3 (CD11b/CD18) in regulating neutrophil adhesion. *J Immunol,* 2002, Dec 1, 169(11), 6417-6426.

Xue, W; Kindzelskii, AL; Todd, RF 3rd; Petty, HR. Physical association of complement receptor type 3 and urokinase-type plasminogen activator receptor in neutrophil membranes. *J Immunol,* 1994, May 1, 152(9), 4630-4640.

Xue, W; Mizukami, I; Todd, RF 3rd; Petty, HR. Urokinase-type plasminogen activator receptors associate with beta1 and beta3 integrins of fibrosarcoma cells: dependence on extracellular matrix components. *Cancer Res,* 1997, May 1, 57(9), 1682-1689.

Ye, F; Hu, G; Taylor, D; Ratnikov, B; Bobkov, AA; McLean, MA; Sligar, SG; Taylor, KA; Ginsberg, MH. Recreation of the terminal events in physiological integrin activation. *J Cell Biol,* 2010, Jan, 11, 188(1), 157-173.

Yebra, M; Parry, GC; Strömblad, S; Mackman, N; Rosenberg, S; Mueller, BM; Cheresh, DA. Requirement of receptor-bound urokinase-type plasminogen activator for integrin alphavbeta5-directed cell migration. *J Biol Chem,* 1996, Nov 15, 271(46), 29393-29399.

Yebra, M; Goretzki, L; Pfeifer, M; Mueller, BM. Urokinase-type plasminogen activator binding to its receptor stimulates tumor cell migration by enhancing integrin-mediated signal transduction. *Exp Cell Res,* 1999, Jul 10, 250(1), 231-240.

Yin, C; Ciruna, B; Solnica-Krezel, L. Convergence and extension movements during vertebrate gastrulation. *Curr Top Dev Biol,* 2009, 89, 163-192.

Zaidel-Bar, R; Itzkovitz, S; Ma'ayan, A; Iyengar, R; Geiger, B. Functional atlas of the integrin adhesome. *Nat Cell Biol,* 2007, Aug, 9(8), 858-867.

Zijlstra, A; Lewis, J; Degryse, B; Stuhlmann, H; Quigley, JP. The inhibition of tumor cell intravasation and subsequent metastasis via regulation of in vivo tumor cell motility by the tetraspanin CD151. *Cancer Cell,* 2008, Mar;13(3), 221-234.

In: Chemotaxis: Types, Clinical Significance... ISBN: 978-1-61728-495-3
Editor: T. C. Williams, pp.53-83

Chapter 2

Plant Growth-Promoting Bacteria: The Role of Chemotaxis in the Association Azospirillum Brasilense-Plant

Raúl O. Pedraza[1], María I. Mentel[1], Alicia L. Ragout[2], Ma. Luisa Xiqui[3], Dulce Ma. Segundo[3] and B. E. Baca [3]

[1]Facultad de Agronomía y Zootecnia, Universidad Nacional de Tucumán. Av. Roca 1900. (4000) Tucumán. Argentina

[2]PROIMI-CONICET. Av. Belgrano y Pje. Caseros. (4000) Tucumán. Argentina

[3] Centro de Investigaciones Microbiológicas, Benemérita Universidad Autónoma de Puebla, Cdad. Universitaria, Edif. 103. Col. San Manuel, 72570. Puebla, Pue. México

Abstract

The genus *Azospirillum* belongs to the plant growth-promoting bacteria group, capable of positively influencing the growth and yield of numerous plant species, many of them with agronomic and ecological importance. Plant growth promotion is largely determined by the efficient colonization of the rhizosphere (soil influenced by roots and microorganisms). Root exudates constitute the most significant source of nutrients in the rhizosphere and seem to participate in the early colonization by inducing the chemotactic response of bacteria. Therefore, chemotaxis is considered an essential mechanism for the successful root colonization by *Azospirillum.*

In this chapter we present a background and new insights on *Azospirillum* chemotaxis, concerning the genetic aspects and its use for addressing biotechnological applications. First, we demonstrate that the genetic complementation of a mutant strain, impaired in surface motility, led to the identification of the gene *chsA* (chemotactic signaling protein). The deduced translation product, ChsA protein, contained a PAS sensory domain and EAL active site domain. The latter has phosphodiesterase activity (PDE-A) for the hydrolysis of c-di-GMP [cyclic-bis (3' –5') dimeric GMP], a compound known to function as a second messenger in different cellular processes, including motility, biofilm formation and cellular differentiation.

After cloning chsA, ChsA protein was expressed and purified by affinity chromatography. ChsA activity in presence of bis-p-nitro phenyl-phosphate was 0.59-ÂµM min-1 mg-1 protein, demonstrating that it displayed phosphodiesterase activity. This suggests that ChsA is a component of the signaling pathway controlling chemotaxis in Azospirillum. Then, we propose that the redox state of the cell is sensed through the PAS domain and directly coupled to the transmitter EAL module, showing PDE-A activity.

Second, the chemotaxis of different strains of A. brasilense toward strawberry root exudates was investigated. The agar-plate assay was used, including two concentrations of exudates from three commercial varieties of strawberry, collected at different time intervals. To quantify the chemotactic response, the capillary method was used. In all cases, a positive chemotactic reaction was found, revealing higher responses in endophytic than in rhizospheric strains, being this strain-specific. Furthermore, the variation of the chemotactic response observed depended on the concentration and time to collect the exudates, as well as the total sugar content. Considering that A. brasilense possessses biotechnological application, addressing to a sustainable agriculture, determining the genes and mechanisms involved in chemotaxis response, as well as the level of activity of strains to root exudates may represent an initial step in selecting them for use as inoculants in different crops.

Introduction

As of today, biofertilizers are considered to be a component of the integrated plant nutrition management. They are defined as substances that contain live microorganisms which, when applied to seeds, roots, plant surfaces or soil, colonize the rhizosphere or inside the plant and promote growth by increasing the availability of nutrients and plant protection in the host (Vessey, 2003). In this context, the plant growth-promoting bacteria are considered to be an alternative to the use of chemical fertilizers and pesticides (Kloepper and Beauchamp, 1992).

At laboratory and field experimental conditions, the effect of biofertilizers has been recognized as a form of sustainable management in agroecosystems (Dobbelaere et al., 2003, Lucy et al., 2004); however, the success of using these biofertilizers lies in the study of compatibility strains and specificity for certain crops and environmental soil conditions.

The large scale use of beneficial microorganisms as biofertilizers in any agricultural production system would bring great benefits because they are cheaper than those of synthetic origin, have positive effects on plants (similar to a chemical fertilizer) and do not exert harmful ecological impacts on the environment or human health.

Most of the bacterial-plant associations occur at the rhizosphere, the soil area that is strongly influenced by plant roots (Vanbleu and Vanderleyden, 2003). These associations are initiated in response to the exchange of signals triggered from the microbe-plant interaction (Vanbleu and Vanderleyden, 2003). The rhizosphere is rich in nutrients, due to the accumulation of a variety of organic compounds released by root through exudation, secretion, and deposition (Curl and Truelove, 1986). These compounds can be used as carbon and energy sources by microorganisms, stimulating the microbial activity, which is particularly intense in the rhizosphere. This was reflected by the number of bacteria found around the roots that are generally 10 to 100 times higher than outside the rhizosphere (Weller and Thomashow, 1994).

The root-associated bacteria capable of stimulating plant growth are generally known as plant growth-promoting rhizobacteria (PGPR) (Davison, 1988; Kloepper *et al.*, 1989). Among them are found species belonging to several genera: *Azotobacter*, *Azospirillum*, *Azoarcus*, *Klebsiella*, *Bacillus*, *Pseudomonas*, *Arthrobacter*, *Enterobacter*, *Burkholderia*, *Serratia* and *Rhizobium*.

The mechanisms by which PGPR promote plant growth are not fully elucidated, but it is known that they include: phytohormones production (Egamberdiyeva, 2007; Shaharoona *et al.*, 2006), biological nitrogen fixation (Mrkovacki and Milic, 2001; Salantur *et al.*, 2006), antagonism against pathogens by production of siderophores, the synthesis of antibiotics, enzymes and/or antifungal compounds (Ahmad *et al.*, 2006; Jeun *et al.*, 2004), and also by the solubilization of phosphates and other nutrients (Cattelan *et al.*, 1999).

In many crops of agronomic importance, significant increases in growth and yield in response to inoculation with PGPR has been reported (Asghar *et al.*, 2002; Bashan *et al.*, 2004; Biswas, *et al.*, 2000). The application of this type of rhizobacteria has resulted in an evident plant growth-promotion, observing an increase in the emergence, vigour, biomass production, root system development and enhancement of up to 30% in the production of commercially important crops, such as sorghum (Raju *et al.*, 1999; Vikram, 2007), tomato (Gravel *et al.*, 2007; Siddiqui and Shaukat, 2002), maize (Kozdroja *et al.*, 2004), wheat (Ozturk *et al.*, 2003; Khalid, 2004), and various grains (Dobbelaere *et al.*, 2001).

The genus *Azospirillum* belongs to the α-subdivision of Protobacteria and has been isolated from the rhizosphere of many plant species in different regions of the world, including tropical and temperate climates. The species belonging to the genus are motile and exhibit chemotaxis to several compound found in roots exudates. The bacterial mobility in the rhizosphere responds to the chemotaxis, which allows them to move towards the roots to obtain benefits from root exudates (source of carbon and nitrogen).

Azospirillum was found in places where the oxygen concentration is optimal for biological nitrogen fixation (Barak *et al.*, 1982); some associate on the root surface and others manage to colonize the root interior (Döbereneir and Pedrosa, 1987), where there is less competition for available substrates, which is very important due to the high demand for energy required for nitrogen fixation (Falk *et al.*, 1985).

Motility provides a survival advantage under a wide variety of environments, allowing bacteria to respond to beneficial or negative conditions and to compete successfully with other microorganisms. *Azospirillum brasilense* possesses a polar flagellum in all culture conditions, and synthesizes lateral flagella when growing on semisolid media (Hall and Krieg, 1983). The flagella of *Azospirillum* are one of the most complex and extremely effective organelle of locomotion, capable of propelling the bacterium through liquids (swimming) and through viscous environments or over surfaces (swarming). In addition, these organelles play an important role in adhesion to substrates and biofilm formation contributing to the interaction with the plant (Croes *et al.* 1993). The colonization of at least part of the root system is required for the beneficial effects of *Azospirillum* inoculants preparations for applications such as biofertilization and phyto-stimulation (Okon and Vanderleyeden 1997).

Many bacteria use a complex behaviour called taxis to sense specific chemicals or environmental conditions and move towards attractants and away from repellents (Adler, 1966). Bacterial taxis is directly involved in interactions with both animal and plant host (Kato *et al.* 2008; de Weert, 2002). Taxis, especially chemotaxis, together with the mechanism of signal transduction and response regulation, have been well studied in *Escherchia coli* and *Salmonella enterica* serovar Typhimurium (Stock and Surette, 1996). Chemotactic ligands are detected by cell surface chemoreceptors called methyl–accepting chemotactic proteins (MCPs). Upon binding a chemotactic ligand, MCPs generate chemotactic signals that are communicated to the flagellar motor via a series of chemotaxis (Che) proteins, *E. coli* possesses 6 Che proteins, one of them, the CheA, is a histidine protein kinase that autophosphorylates at a specific histidine residue to form CheA~P. The phosphoryl group of CheA~P is transferred to a specific aspartate residue of CheY, to form activated CheY~P which is a response regulator of a two component regulatory system, that interacts directly with the flagellar motor switch protein to control the direction of flagellar rotation. CheZ acts as a negative regulator involved in inactivation of CheY~P by dephosphorylation of CheY~P to CheY. MCPs, with help from CheW, modulate the autophosphorylation activity of CheA in response to temporal changes in stimuli intensity. MCPs undergo reversible methylation at several glutamate residues mediated by CheR and CheB that are methyltransferase and methyl esterase enzymes, respectively. CheB is another response regulator, which is phosphorylated by CheA~P to form CheB~P. CheA~P exhibits higher methylesterase activity than CheB and is not affected by environmental stimuli. CheR continually adds methyl groups to MCPs, and constitutes the via of adaptation response. The methylation level of MCPs is controlled in response to environmental stimuli and affects their conformation. This reversible methylation of MCPs is required for temporal sensing of chemical gradients (Stock and Surette, 1996).

Pedraza *et al.* (2007) demonstrated the natural occurrence of *A. brasilense* colonizing strawberry plants, including inner tissues of roots and stolons. The latter would provide an additional agronomic advantage, considering the asexual propagation of those plants in commercial nurseries, by stolon fixation on the ground. Thus, if the strawberry plants are inoculated with *Azospirillum* strains selected by their PGPR characteristics, the presence of this bacterium could be insured in their descendants.

From the results of inoculating strawberry plants with different strains of *Azospirillum*, and considering that the best plant growth depends on the interaction between specific genotypes of bacteria and plants (Pedraza *et al.* 2009), the use of this genus represents an interesting option for higher agricultural production and significant ecological advantages. Hence to increase the knowledge of the *Azospirillum*-plant interaction, in the present chapter we consider the chemotaxis as a starting point for a successful partnership.

Physiology of *Azospirillum* Chemotaxis

It has been demonstrated that *A. brasilense* polar flagellum rotates in both clockwise and counterclockwise directions. The last one rotation causes forward movement of free-swimming cells, while the change in the direction of rotation to clockwise cause a reversal in swimming direction. When *Azospirillum* cells are exposed to malate, a strong attractant, some effects in swimming behaviour are observed: suppression of direction change, the chemotaxis, and a long term-increase in swimming speed as chemokinesis (Zhulin and Armitage, 1993). In fact, *Azospirillum* strains responded chemotactically to temporal gradient of some effectors such as amino acids, sugars and organic acids, as well as to maize mucilage. This behaviour is strain dependent, suggesting that a certain degree of specificity exists in the establishment of plant-bacteria interaction (Reinhold *et al.* 1985; Mandimba, *et al.* 1986). The presence of oxidizable substrates increased the number of attracted bacteria only 1.2 to 3-folds, a rather low ratio, compared with other bacteria. Further work revealed that the attraction was to oxygen gradient dissolved in water, and the exposure of the cells to an oxygen gradient followed by aerotaxis, masked the chemotactic response of the bacteria (Barak *et al.* 1982). Therefore, aerotaxis is an important response in *A. brasilense,* which guide the bacterial to a preferred low oxygen concentration for energy generation (3 to 5 μM). Indeed, the proton motive force was lower at oxygen concentrations that were higher or lower than the preferred oxygen concentration. It was suggested for *A. brasilense* that to reach the optimal oxygen concentration was relevant for energy generation and nitrogen fixation in the rhizosphere (Zhulin *et al.* 1996).

It was described that there are several ways to sense chemicals: Chemotaxis usually is referred as to metabolism-independent behaviour, while the use for "energy taxis" denotes metabolism-dependent behaviour. Energy taxis is broadly defined as a behavioural response to stimuli that affect cellular energy levels. It includes responses directly linked to electron transport/energy generation, such as aerotaxis, redotaxis and phototaxis (Taylor and Zhulin, 1998). The most important behaviour in *A. brasilense* is energy taxis and it was demonstrated that the compounds which are attractants for *Azospirillum* cells are metabolizable subtracts, while their nonmetabolizable analogues are not attractants. On the other hand, the inhibition of the metabolism of a chemical attractant completely abolishes chemotaxis to this compound. Moreover, it was observed the correlation between the efficiency of a chemical as a growth substrate and as a chemoeffector (Alexandre *et al,* 2000). Chemicals that interact directly as inhibitors of electron transport were found to be strong repellents, and most important, a mutant lacking the cytochrome ccb_3-type terminal oxidase had significantly diminished chemotaxis to all major attractants, but only under microaerobic conditions. When it was assayed under fully aerobic conditions, where this respiratory system is not functional, chemotactic responses in the mutant and wild-type strain were identical. The results showed that the signal for chemotaxis toward major attractants and repellents is originated within a functional electron transport system in *A. brasilense* (Alexandre *et al,* 2000).

The Research for the Chemoreceptors in *Azospirillum*

Characterization of chemoreceptors capable of measuring changes within the electron transport system is required in order to conclusively establish sensing mechanisms. The search for plant-inducible bacterial genes from *A. brasilense* led to identification of a 40kDa protein, which was induced in presence of wheat root exudates. The protein was proteolytic cleavage and the sequence of two peptides was used to obtain the corresponding gene, named *sbpA*. The cloning, sequencing and bioinformatics analysis of the coding DNA region revealed significant homology with ChvE protein from *Agrobacterium tumefaciens.* ChvE is a periplasmic sugar-binding protein, also functions in the uptake of sugars and chemotaxis of *A. tumefaciens* towards sugars (Van Bastelaere *et al.* 1999). Further, it was determined the role of SbpA in chemotaxis activity towards D-galactose, L-arabinose and D-fucose. It was interesting to note that the response of *A. brasilense* to sugars was inducible. This was confirmed by the expression analysis of a transductional fusion *sbpA::gusA*. D-galactose, L-arabinose and D-fucose strongly induced the gene expression. Furthermore, the mutant $sbpA::km^R$ was severely affected in uptake of D-galactose. Then, SbpA is part of the binding protein-dependent, a high affinity uptake system, and is required for chemotaxis towards D-galactose. Additionally, the chemosensory pathway seems to be dependent on the uptake and metabolism of the attractant (Van Bastelaere *et al.* 1999).

A protein acting as energy taxis chemoreceptor named Tlp1 (for transducer-like protein 1) was identified in *A. brasilense* Sp7 strain. The protein structure comprises the functional domains characteristics of chemoreceptors. It has a membrane topology typical of classical membrane chemoreceptors, the N-terminal periplasmic sensing domain, and a C-terminal region which consists of a HAMP (histidine kinase, adenylcyclase, methyl binding proteins, and phosphate) domain and a signalling module containing the HCD (highly conserved domain) and two methylation regions typical of chemosensors (Greer-Phillips, *et al.* 2004). The gene mutated *tlp1* exhibited a phenotype deficient in chemotaxis to several oxidizables subtracts, taxis to terminal electron acceptor such as oxygen and nitrate, and redox taxis to substituted quinones. Altogether, suggested that Tlp1 mediates energy taxis in *A. brasilense.* Furthermore, the *tlp1* mutant was severely affected in colonization to wheat root as defined by qualitative and quantitative ß-galactosidase assays. This indicated that Tlp1 protein was acting as chemoreceptor guiding the bacterium by allowing it to locate and navigate towards a habitat optimal for growth. Considering that root exudates could serve as growth substrates, the energy taxis might be involved in root colonization and the establishment of *Azospirillum*-grass associations (Greer-Phillips, *et al.* 2004).

Genetics and Biochemical of Determinants Involved in Chemotaxis in *Azospirillum*

The identification of genes encoding by the excitation (CheA, CheW and CheY) and adaptation (CheB and CheR) chemotaxis pathways from *A. brasilense* was obtained by genetic complementation of two generally non-chemotactic mutants. The genes identified showed high identities (>50%) with the corresponding genes to α-proteobacterias, and were located in tandem, suggesting an operon structure (Hauwaerts *et al.* 2000). Although a previous work has showed that the chemotactic response of *A. brasilense* to most strong attractant malate has been methylation independent (Zhulin & Armitage, 1993), the presence of the *cheR* (coding for a methyl esterarse), and *cheB* (coding for a methyltransferase) genes, indicates that responses to at least some stimuli requires methylation and demethylation of the chemotaxis transducers (Hauwaerts *et al.* 2000).

Further work revealed that CheR, CheB and CheBR contributed to chemotaxis and aerotaxis significantly in *A. brasilense* but they are not essential for these behaviors to occur (Stephens *et al.* 2006). The chemotacting rings formed by the corresponding mutants were significantly smaller than those formed by the wild-strain for the strong attractants tested (malate succinate, fructose). Using the spatial gradient assay *cheB* and *cheR* mutant were unable to form aerotaxis bands. However the *cheRB* mutant formed band, but farther away from the meniscus than the one formed by the wild-type. None of the three mutants responded to either addition or removal of oxygen for aerotaxis, whereas the wild-type had a positive and negative response, respectively. The liberation profile of methanol, that is a measure of methyl estearase (CheB) activity (involved in the adaptive response) was very different with the attractants tested to those obtained with *E. coli.* Interestingly, the wild-type strain on the addition or removal of oxygen and succinate did not induce the production of methanol, indicating that in this bacterium there is a methylation-independent pathway for those molecules, confirming the early results obtained (Zhulin & Armitage, 1993). Furthermore, the *cheBR* mutant release methanol upon both in addition or removal of succinate, suggesting that there are more than one methylation/demethylation system(s) in *A. brasilense.* In order to test this hypothesis the authors, constructed a Δ*cheA::gusA-km* mutant. The resulting *che* operon mutant was impaired in, but not null for chemotaxis. Taking together all this data strongly suggesting that: i).The operon under study is partially involved in aerotaxis and quimiotaxis, but it is not its principal function. ii). The aerotaxis in *A. brasilense* appears to be methylation independent. iii).The data strongly suggested that in *A. brasilense* occur several quimiotaxis pathways (Stephens *et al.* 2006). Analysis of the draft genome sequence of *A. brasilense* Sp245 strain suggests the presence of four *che*-like signal transduction pathways, multiple homologous of adaptation proteins and about 30 chemotaxis transducers (Alexandre. 2007).

Identification of Chsa, A Novel Protein Involved in Chemotaxis

It was recently obtained the characterization of a mutant strain of *A. brasilense* Sp7S impaired in surface motility and chemotactic response (Carreño-López *et al.* 2009). The genetic complementation of Sp7S strain (Fig 1) and the nucleotide sequencing of the complementing region conducted to the identification of *chsA* gene coding for a putative signal transduction protein designated ChsA. The nucleotide sequence of the *chsA* region has been deposited in the GenBank database with accession number AM408892.

The inferred ChsA protein (63,714 Da, 586 residues) has characteristics similar to cytoplasmic signalling proteins. It contains two domains: a PAS sensory domain near the N terminus region (from residues 50 to 160) and an EAL transmitter domain in the C-terminus part (from residues 330 to 560). The EAL domain is present in proteins with phosphodiesterase activity (PDE-A) involved in the hydrolysis of c-di-GMP (cyclic-bis (3'-5') dimeric GMP), a compound known to function as a second messenger (Figure 1) in a broad spectrum of cellular processes including motility, biofilm formation and cellular differentiation (Jonas *et al.* 2009). Proteins containing an EAL domain often possess another domain GGDEF, carrying guanylate cyclase activity, but this domain is not present in ChsA. The PAS sensory domain is found in a variety of proteins with redox functions, including NifL, Dos and Aer, which typically bind heme, and flavines (Taylor and Zhulin 1999; Greer-Phillips *et al* 2003). Databases searches using the BLASTP program revealed that ChsA shared similarity with a limited number of uncharacterized ORFs from α Proteobacteria, phylogenetically close to *Azospirillum*, such as *Magnetospirillum magnetotacticum* (YP_421664.1) and *Rhodospirillum rubrum* (YP_426135.1), (Carreño-López *et al.* 2009).

It was constructed the *chsA::km*^{R} mutant and its phenotype determined. The presence of flagella in the mutant strain was confirmed by Western blot analysis, using antisera raised against the polar and lateral flagellins, and by electron microscopy. It displayed the same growth rate and motility as the wild type in liquid medium; however, showing a defect in motility in semi-solid minimal medium added with several metabolizable subtracts, defined as strong chemoattractants for *Azospirillum,* such as: to three organic acids, malate, succinate and pyruvate and the amino acids glutamate and proline. It was employed as a negative control a mutant lacking the sigma N factor (σ^{54}); the *rpoN* gene controls the expression of both flagella in *Azospirillum* (Milcamps *et al.* 1996). As expected, strain *rpoN* was completely non-chemotactic. It was observed a significant decrease (c.a. of 20 %) in the chemotactic response with all the substrates used, in the case of *chsA*-Tn*5* mutant strain (Sp74031) as compared to the wild type. A partially reduced chemotactic response was also reported for *cheB* and *cheR* mutants for the adaptation pathway, suggesting multiple chemotaxis systems in *Azospirillum* (Stephens *et al.* 2006; Bible *et al.* 2008)

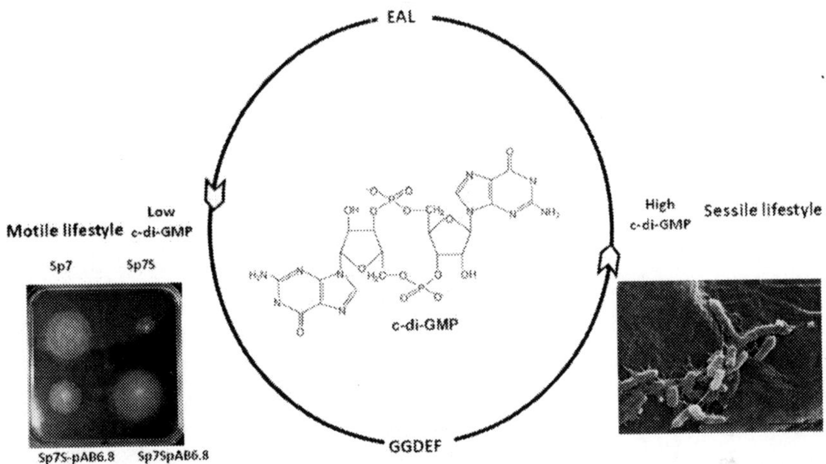

Figure 1. The role of c-diGMP in bacterial sessile and motile lifestyle. In *A. brasilense* Sp7, regulates swimming motility (left, genetic complementation of Sp7S mutant), and biofilm formation (right) in response to changing cellular pool of c-di-.GMP. On left an image of genetic complementation of *A. brasilense* Sp7S strain mutant by the plasmid pAB6.8, which carried the gene *chsA*. The strains were grown on K-malate minimal medium with 0.25% agar for 48h. On the right is a SEM of *A. brasilense* Sp7 biofilm

In addition, it was analyzed the enzymatic activity of ChsA as an avenue to gain information about the functional characteristics of this novel chemotaxis signal transduction protein, and to expand our knowledge on the behavior of an EAL protein. The *chsA* gene was cloned, the protein expressed and purified for the determination of phosphodiesterase activity. For that, the genomic DNA of *A. brasilense* Sp7 was isolated following standard procedures and the gene was amplified by PCR. The amplicon was cloned in the expression vector pBAD (Invitrogen). The plasmid harboring the gene and the C-terminal His_6 tag-encoding sequence were transformed into *E. coli* strain BL21(DE3). The expression results are presented in Figure 2.

The protein purification was done from 500ml bacterial culture (LB medium), which was grown to an optical density of 0.8 (OD_{600} nm) before being induced with 0.08% L-arabinose. The culture was shaken at 16°C for 12 h before being pelleted by centrifugation. The cells were lysed in 20 ml lysis buffer (20 mM Tris [pH 8.0], 500 mM NaCl, 5% glycerol, 0.1% β-mercaptoethanol, 0.1% Triton X-100, and 1 mM phenylmethylsulfonyl fluoride). After centrifugation at 25,000g for 30 min, the supernatant was filtered and then incubated with 2 ml of Ni_2-nitrilotriacetic acid resin (Qiagen) for 1 h at 4°C. The resin was washed with 50 ml of W1 buffer (lysis buffer with 20 mM imidazole) and 20 ml of W2 buffer (lysis buffer with 50 mM imidazole). The proteins were eluted using a stepped gradient method with the elution buffer containing 20 mM Tris (pH 8.0), 500 mM NaCl, 5% glycerol, and 200 mM, 300 mM, or 500 mM imidazole. After sodium dodecyl sulfate-polyacrylamide gel electrophoresis analysis, fractions with purity higher than 95% were pooled, and the enzymatic activity determined (Figure 3).

Using a purified his-tag fusion of ChsA, it was demonstrated that this protein has phosphodiesterase (PDE-A) activity in the presence of Mn^{2+} against the artificial substrate bis(*p*-nitrophenyl) phosphate (bis-*p*NPP). This *in vitro* activity was considerably higher at 37°C than at room temperature. In this study we have determined that *in vitro* the pH optimum for the PDE activity of EAL-ChsA was between pH 8 and 9 (Figure 3). Since the proposed function of EAL-domain proteins is to linearize c-di-GMP (Schmidt *et al.* 2005; Schimer and Jenal, 2009), this is a direct demonstration of the required phosphodiesterase activity of ChsA purified EAL-domain protein, which likely degrades the c-di-GMP.

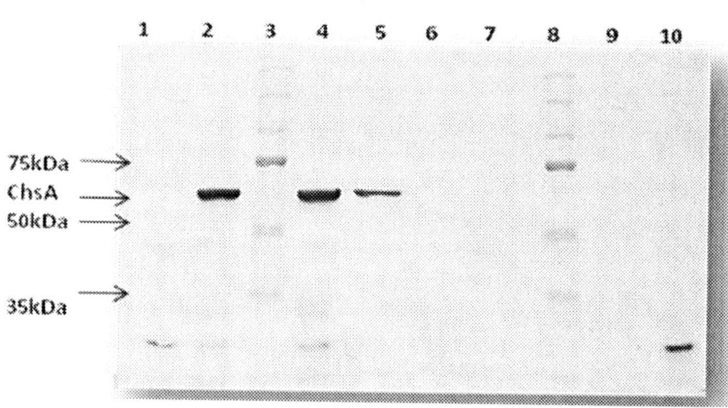

Figure 2. Expression of ChsA in *E. coli* BL21(DE3) strain. Transformed BL21(DE3) cells were grown at 30°C for 3h. The expression was performed in the presence of 0.08 % of L-arabinose at 16°C for 5 h post-induction. Expression from each sample was tested with 50µg of protein. Representative Western blot probed with an anti-His antibody showing the expression of the C-terminal His_6 ChsA after induction of expression, as well as the cellular localization of ChsA. Lane 1. *E. coli* pBAD (vector), pellet; lane 2. ChsA (pellet); lane 3. M.M.; lane 4. ChsA (pellet); lane 5. ChsA (soluble fraction); lane 6. *E. coli* pBAD (vector, soluble fraction; lane7. empty; lane 7 M.M; lane 10 pBAB (vector, pellet)

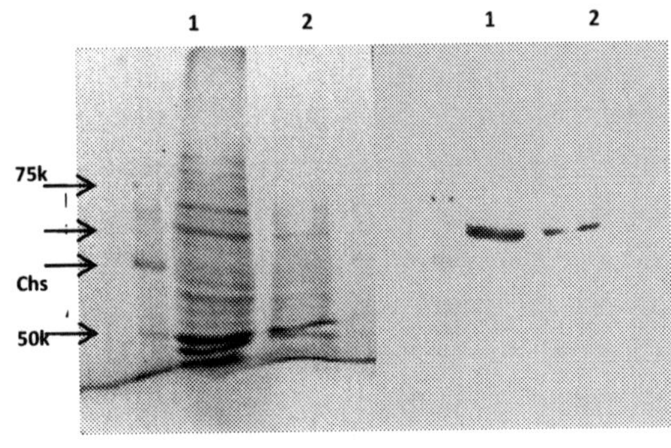

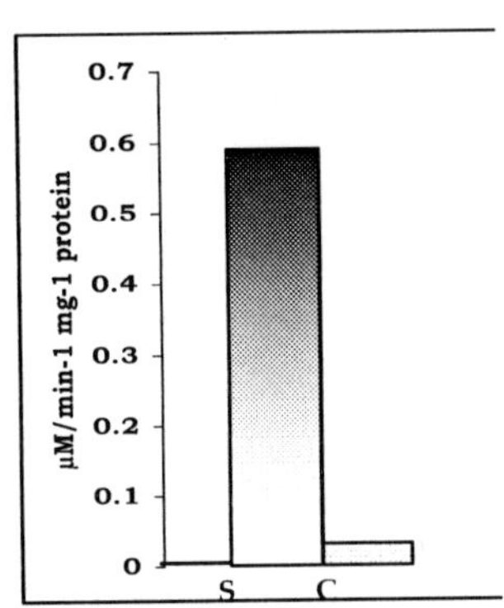

Figure 3. Purification and activity of recombinant ChsA by affinity chromatography. (a) Protein was detected by Western blotting of purified His-tag-ChsA, using antiserum against His6 (Sigma chemical Co. USA) (b). 10 µg of affinity purified, dialyzed ChsA was tested for PDE activity in a reaction buffer (0.05 M Tris, 0.05 M Bis-Tris, 0.1M Na acetate, 1mM MnCl2, 2.5 mM of bis-pNPP) at pH 8.5, as previously described by Bobrov et al. (2005); S = Sample, C = Control

It is worth noting to find that others genes such as *chsA* might contribute to chemotaxis response, as revealed for this gene. Many of the proteins involved in the metabolism of the second messenger compound c-di-GMP contain both EAL and GGDEF domains, responsible respectively for phosphodiesterase and diguanylate cyclase activities (Schmidt *et al.* 2005; Schimer and Jenal, 2009), however, ChsA only contains the EAL domain. It has been shown that the YhjH and EAL domain proteins with phosphodiesterase activity from *S. enterica* sv Typhymurium are implicated in the transition between sessility to motility (Simm *et al.* 2004). Surveys of bacterial genomes revealed that the number of genes encoding proteins with GGDEF or EAL domains could vary from none to up to 99, suggesting that the role of each individual protein may be extremely complex and that their inactivation is unlikely to result in clear-cut phenotype in case of high multiplicity (Galperin 2005). Deduced translation of *chsA* indicated that ChsA product contains a PAS domain. Proteins containing a PAS domain are present in numerous bacteria and are involved in sensing a variety of environmental signals (Taylor and Zhulin 1999). It is tempting to speculate that the PAS domain in ChsA protein senses the redox state of the cell through a cytoplasmic signaling molecule directly coupled to the transmitter EAL module. Here, we presented biochemical data to demonstrate that the EAL domain of ChsA is catalytically active, with phosphodiesterase activity.

However, the relationship between ChsA and the regulation of the chemeotactic machinery still remains unknown.

The Role of Bacterial Motility and Chemotaxis in *Azospirillum*-Plant Interaction

It has been reported that certain strains of *A. brasilense* show positive chemotaxis towards various attractants such as sugars, amino acids, organic acids (Okon *et al.*, 1980; Barak *et al.*, 1982; Reinhold *et al.*, 1985; Zhulin and Armitage, 1992), as well as to root exudates (Heinrich and Hess, 1985; Mandiambra *et al.*, 1986; Okon *et al.*, 1980; Bacilio-Jiménez *et al.*, 2003; Pedraza *et al.* 2009). This mobility is important as it allows the bacterial access to a more suitable habitat and become more competitive with other microorganisms in the root.

In natural environments, soil moisture is a limiting factor in the migration of *A. brasilense* to the roots of plants (Bashan, 1986). This suggests that the "swimming movement" plays an important role in such environments. The ability of *A. brasilense* to start the root colonization of wheat was investigated with different mutant strains (Vande Broek *et al.*, 1998). Only non-flagellate and non-chemostatic mutants showed a strong reduction in the capability for colonization, which showed the requirement of bacterial mobility to initiate root colonization of wheat (Vande Broek *et al.*, 1998).

Bashan and Holguin (1994) compared the interroot movement of *A. brasilense* wild-type and its isogenic non-motile strain, which were inoculated to the root systems of soybean and wheat seedling, in presence of chemoattractants and repellents. They showed highly differences between both strains, no movement of *A. brasilense* Mot$^-$ cells from inoculated roots was detected. The inoculated *A. brasilense* Mot$^-$ bacteria remained at the seed inoculation and did not migrate with the root tips; whereas wild-type strain showed a differential pattern, Mot$^+$ cells migrated for several centimetres from inoculated to non-inoculated roots, irrespectively of plants species. Indicating, that motility and chemotaxis are actives processes involved in efficient colonization (Bashan and Holguin, 1994)

By a different approach using the ß-galactosidase activity as a reporter system it was showed that only motility and chemotactic mutants from *A. brasilense* Sp7 wild-type strain were affected in their capacity to initiated wheat root colonization at the root hair zones. Indeed the ß-galactosidase activity determined by qualitatively and quantitatively assays from NM313 mutant strain (altered in chemotaxis to succinate and citrate), and Sp7p90Δ84 strain (a non-flagellated mutant; van Rhijn *et al.* 1990), differ significantly from Sp7wild-type strain in colonization to root wheat. This indicates that chemotactic movement is important in the initiation of wheat root colonization (Vande-Broek, 1998).

In the rhizosphere, it may be advantageous to respond positively to any compound that increases metabolic rates, such as organic acids and oxygen. Low oxygen concentrations typically of the rhizosphere are seen to be one of the main stimuli that attract bacteria to plant roots. Aerotaxis in the rhizosphere would be expected to provide a real advantage to bacteria when searching for nutrients and in competition with other microorganisms (Barak *et al.*1982; Zhulin and Armitage, 1993).

Chemotaxis of *A. Brasilense* towards Strawberry Root Exudates Assessed by the Agarose-Plate Method

Strawberry *in vitro* plants of three commercial varieties ('Camarosa', 'Milsei' and 'Selva') were aseptically grown in 25 ml of diluted (1:2) Hoagland solution (Hoagland 1975) and maintained in a growth-chamber at 25°C, 70% of relative humidity and a photoperiod of 16 h light. The root exudates were collected from the liquid nutrient medium (Hoagland solution) used by plants after 7, 14, and 28 days of growth. The nutrient medium (25 ml) containing the root exudates was removed and sterilized by filtration (0.2 µm Millipore), lyophilized and kept at -20°C for total protein and sugar determination and chemotaxis test. Sterility of each solution was verified by plating samples in LB medium and incubated 72 h at 30°C.

Chemotaxis was evaluated on SM medium (Reinhold et al. 1985), without malic acid, yeast extract, neither NH_4Cl, and supplemented with 0.3% agarose (w/v). Lyophilized extracts containing root exudates from each cultivar were resuspended in distilled sterile water to obtain two concentrations: 8x (d1) and 4x (d2). Root exudates (0.1 ml) was added to 7 ml of SM medium (kept at 45°C), vigorously mixed and poured into sterile Petri dishes (60 mm diameter). Once at room temperature 0.01 ml of previously grown and washed bacteria (see below) was placed in the centre of the plates. Plates were incubated at 30°C and the halo diameter measured (mm^2) after 48 h. Mobile bacteria were obtained from 48 h old cultures grown in SM medium (Reinhold et al. 1985). Cells were collected by centrifugation (15 min at 15,000xg) and washed three times with potassium phosphate (60 mM pH 7.0)/Na–EDTA (0.1

mM) buffer. The cells were finally resuspended in phosphate buffer without EDTA and the concentration adjusted to 10^8 cells ml^{-1} (OD_{600} = 1.0). Cell motility was controlled by contrast phase microscopy (Olympus BH-2). For chemotaxis test, cell suspensions were used within 1 h after washing to avoid motility loss.

Chemotaxis assays were carried out on a complete randomized factorial design, including four factors: bacterial strains, strawberry cultivars, root exudates dilution (d1, d2), and the time the root exudates were collected (t1, t2, t3). ANOVA was performed, and the main effect of the different treatments was evaluated by the Wald Test ($P \leq 0.05$), using the software Infostat 2.0.

The positive chemotactic response was visualized by halo formation on culture medium surface. Table 1 shows the Wald test results conducted from data obtained in measuring the surface of the halos formed. These show that the main effects of factor variety ($p < 0.01$) and the dilution factor ($p < 0.05$) were statistically significant, as well as the interaction between bacterial strain and strawberry variety ($p < 0.05$), as indicating that the chemotactic response of the strain depends on the variety of the plant. One factor affecting the size of the halo was the consistency of the agarose, therefore, the addition of Tween 80, recommended by Niu *et al.* (2005) decreased the tension surface, thus favouring the displacement motion by microorganisms and the formation of larger, regular-shape and easy to measure halos.

Table 1. Test of significance for the coefficient model of main effects and interaction of two factors on the response of halo formation

1	F. D.	Wald test	P
Coefficient	1	235.5059	0.000
Strain	3	6.8742	0.070
Variety	2	12.2044	0.002
Strain*Variety	6	13.3457	0.040
Time	2	3.2217	0.200
Variety*Time	4	2.8608	0.600
Strain*Time	6	4.5185	0.600
Dilution	1	5.3759	0.020
Time*Dilution	2	1.5212	0.500
Variety*Dilution	2	0.6355	0.700
Strain*Dilution	3	0.4640	0.900

F.D.: fredon degree; P:probability.

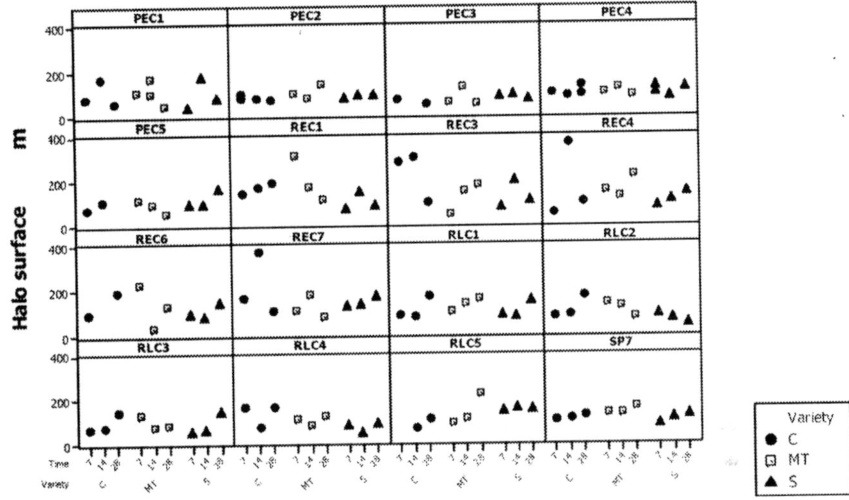

Figure 4 . Time effect of root exudates collection in different strawberry varieties on 16 strains of *A. brasilense* using a 2x exudates concentration

Once the optimal concentration of exudates for chemotaxis assay on agarose plates was determined, 15 strains, characterized in a previous work (Pedraza *et al.*, 2007), were assessed: REC1, REC3, REC4, REC6, REC7, PEC1, PEC2, PEC3, PEC4, PEC5, RLC1, RLC2, RLC3, RLC4, RLC5, and the strain Sp7 as a control. With them, it was determined the chemotactic response towards root exudates in a 2x concentration of each variety of strawberry, taken during three periods, with the aim of selecting strains with high and low chemotactic response to be evaluated quantitatively.

Figure 4 shows the effect of time of collection of root exudates of different strawberry varieties in the response of different strains of *Azospirillum brasilense*. In all cases, we observed a higher number of positive responses to root exudates taken at 7 and 14 days. Endophytic strains (REC1, REC3, REC4, REC6 and REC7), isolated from sterile roots, showed a higher chemotactic response with formation of halos with areas greater than 200 mm^2. The lower chemotactic activity was observed in strains isolated from the rhizosphere and the first stolon (RLC2, RLC3, RLC4, PEC2, PEC3 and PEC4), which showed in all cases halos surfaces less than 200 mm^2. The best chemotactic response was obtained with strain REC3, which showed halos grater than 200 mm^2 for root exudates taken at 7 and 14 days of plant growth, from variety Camarosa. The strain RLC3 showed halos formation with surfaces smaller than 200 mm^2, being the one with less chemotactic response towards the root exudates of the three strawberry varieties collected at different times.

Chemotaxis of *A. Brasilense* Towards Strawberry Root Exudates Assessed by the Capillary Method

From the results obtained in testing agarose-plates, three bacterial strains were chosen to quantify the chemotactic response by capillary method and determine the R_{chem} (chemotaxis index) of them: REC3 as the strain of higher chemotactic response, RLC3 as the lowest response, and Sp7 as control. The quantification of the chemotactic effect was performed by using the capillary method according to Guocheng and Cooney (1993). For that, a sterile plastic chamber with multiple wells was used. The capillaries (1μl of capacity) were placed in the centre of each well and filled using a sterile syringe. To observe the filling of capillaries, the attractant solution was mixed with Comassie blue. One end of the capillary carrying the attractant solution was introduced through the central hole silicone septa in the cell suspension (250 μl) contained in the chamber well. The free end of the capillary was sealed with plastic putty to avoid the aerotaxis effect. The chamber was incubated during one hour at 30°C. Then, the content of the capillary were expelled using a sterile syringe into microtubes (1.5 ml) with 1ml of chemotaxis buffer and serial dilutions were made (10^{-3}, 10^{-4}, 10^{-5}). Dilutions were plated on solid NFb medium (described in Pedraza *et al.* 2007). Each strain was processed with a capillary control that contained chemotaxis buffer without the attractant solution. The plates were incubated for 72 h at 30°C. After that time the CFU/ml was determined. The ratio between the number of bacteria in the capillary due to chemotaxis and the number due to random motion is known as R_{chem} (chemotaxis index), representing the chemotactic activity of the microorganism towards a particular substance (Barbour *et al.*, 1991). Results are shown in Table 2 and they show concordance with those obtained on agarose-plate method.

The strain REC3 showed the highest chemotactic activity with exudates from the three varieties of plants taken after 7 days of growth. The highest values were obtained with the variety 'Camarosa' (11.98 ± 2.39), followed by the variety 'Milsei' (6.33 ± 0.59) and 'Selva' (3.11 ± 0, 45). These results agree with those obtained by the plate method. The minimum value obtained for this strain was 1.47 with exudates of the variety 'Selva' taken at 28 days.

Table 2. Chemotactic response of *A. brasilense* strains towards root exudates of three varieties of strawberry: Camarosa (C), Milsei (M) and Selva (S), collected in three periods: t7, t14 and t28 (7, 14, 28 days of plant growing), assessed by the capillary method (R_{chem}: chemotactic index; SD: standard deviation)

Strain-Strawberry var.-Time of collecting root exudates	Chemotactic response
	$R_{chem} \pm SD$
REC3 M t7	6,33 ± 0,59
REC3 M t14	2,82 ± 1,53
REC3 M t28	3,97 ± 0,41
REC3 C t7	11,98 ± 2,39
REC3 C t14	6,69 ± 3,22
REC3 C t28	3,67 ± 2,62
REC3 S t7	3,11 ± 0,45
REC3 S t14	4,51 ± 1,52
REC3 S t28	1,47 ± 0,00
RLC3 M t7	0,90 ± 0,39
RLC3 M t14	1,18 ± 0,17
RLC3 M t28	2,06 ± 0,06
RLC3 C t7	0,00
RLC3 C t14	0,93 ± 0,11
RLC3 C t28	1,37 ± 0,89
RLC3 S t7	0,75 ± 1,06
RLC3 S t14	0,55 ± 0,05
RLC3 S t28	0,70 ± 0,28
Sp7 M t7	0,00
Sp7 M t14	0,52 ± 0,05
Sp7 M t28	0,58 ± 0,07
Sp7 C t7	0,00
Sp7 C t14	0,47 ± 0,20
Sp7 C t28	1,60 ± 1,29
Sp7 S t7	2,28 ± 1,01
Sp7 S t14	0,65 ± 0,44
Sp7 S t28	1,55 ± 1,48

The strain RLC3 showed a maximum R_{chem} (1.37 ± 0.89) for the variety 'Camarosa' at 28 days as in the previous test; this value was lower than the value obtained with strain control Sp7 (1.60 ± 1.29). Chemotactic response was not detected by this method when using exudates collected from variety 'Camarosa' 7 days growth, as well as the control strain Sp7 with exudates from 'Milsei'. The results determined by the capillary method confirmed that the chemotactic activity depends on the bacterial strain, plant variety, and time of collection of the exudates. Although the chemotactic response towards strawberry root exudates was performed through two approaches: the method of plates and the quantitative method of capillaries, similar results were obtained in both cases.

Sugars and Proteins Determination of Strawberry Root Exudates

To explain the chemotactic behaviour previously observed, total sugars and proteins content of root exudates were determined. Total sugars was determined from lyophilized samples of root exudates suspended in sterile distilled water as indicated in Pedraza *et al.* (2009). Briefly, 0.64 ml of phenol 80% and 2.5 ml of concentrated H_2SO_4 (80%) was added to 1 ml of each sample, mixed for 1 min with vortex and kept at 30°C for 10 min. After colour development OD_{490} was measured and sugar content evaluated with a standard curve made with different glucose concentrations. Three determinations were performed for each sample (root exudates and glucose standards).

The major amount of sugars was detected in the root exudates obtained at 7 days of plant growing of the three cultivars ('Milsei', 'Selva', 'Camarosa'), then, a diminution of them was observed at 14 and 28 days. The amount of total sugars varied among plant cultivars, being the cv 'Milsei' the highest producer at 7, 14 and 28 days as compared with the cvs 'Camarosa' and 'Selva'. The maximum concentration of sugar (0.0806 mg/ml) was found in root exudates of the variety 'Milsei' taken at 7 days of plant growth, followed by exudates from the variety 'Selva' (0.0629 mg/ml), taken in the same period, and finally in the variety 'Camarosa' (0.0466 mg/ml). The minimum value of sugar concentration (0.0219 mg/ml) was determined in root exudates taken at 28 days from variety 'Camarosa'. The variety 'Selva' showed its lowest concentration (0.0245 mg/ml) in exudates obtained at 14 days and in the variety 'Milsei' the lowest value (0.0358 mg/ml) was detected at 28 days of plant growth.

In contrast, protein concentration of the root exudates, determined according to Bradford (1976) using bovine-serum albumin as standard, was detected only after 14 and 28 days of plant growth. At 28 days, in 'Camarosa' it was determined the maximum value (0.030 mg/ml), followed by the variety 'Milsei' (0.024 mg/ml) and finally the variety 'Selva' with a concentration of 0.019 mg/ml. The minimum value detected was 0.018 mg/ml in the variety 'Selva', then 'Milsei' with a value of 0.024 mg/ml and finally 'Camarosa' (0.025 mg/ml), all detected at 14 days of plant growth.

Conclusion

Azospirillum as motile genus bacteria are capable of navigating in gradients of oxygen, redox molecules, and nutrients by constantly monitoring their environment in order to inhabit where it is optimal for survival and growth. Although there is no strict host specificity in *Azospirillum*-plant associations, a strain-specific chemotaxis was reported. Strains isolated from the rizosphere of a particular plant demonstrated preferential chemotaxis towards chemicals found in root exudates of that plant. These results suggested that chemotaxis may contribute to host-plant specificity, and could largely be determined by metabolism.

Until now, it has appeared that the genetics and molecular traits determined in *Azospirillum* were complex enough. It has been reported that *Azospirillum* undergoes methylation-dependent and independent chemotaxis. Not all chemotactic responses require a methylation-dependent. For example, in *E. coli*, aerotaxis and chemotaxis to phosphoenolpyruvate transport system sugars are methylation independent (Alexandre & Zhulin, 2001). It has been shown that quimiotaxis for such as succinic acid and oxygen are methylation-independent (Stephen *et al.* 2006). On the other hand, the genome sequence of *A. brasilense* Sp245 strain under progress has showed to contain several *che*-like genes, which could be probably involved in the chemotactic behaviour.

C-di-GMP is an intracellular signaling molecule that has been proposed to control the transition between biofilm and planktonic mode of growth (Römling *et al.* 2005). High intracellular c-di-GMP promotes the production of biofilms, adhesives organelles such as pili or stalks, and reduces motility in a certain number of species (Hickman *et al.* 2005; Jonas *et al.* 2009). The analysis of completely sequenced bacterial genomes has revealed that some bacterial species have multiples homologous chemotaxis-like signal transduction pathways (Szurmant and Ordal, 2004). Interestingly, Che-like pathways have been recently implicated in controlling cellular functions other than motility, including flagellum biosynthesis and biofilm formation (Hickman *et al.* 2005; Bible *et al.* 2008). Indeed, a chemosensory system that regulates biofilm formation through modulation of cyclic c-di-GMP has been described in *Pseudomonas aeruginosa*. The Wsp system includes a predicted membrane-bound [methyl-accepting chemotaxis protein (MCP)]-like receptor (WspA), CheW-like scaffolding proteins (WspB and WspD), a CheA/Y hybrid histidine sensor kinase with a received domain (WspE), a methyltransferase CheB homologue (WspC), a methylesterase CheB homologue (WspF), and a two domain response regulator GGDEF protein WspR, which catalyzes the synthesis of c-di-GMP. The system stimulates production of biofilm and suppresses motility, by the formation of c-di-GMP intracellular levels (Hickman *et al.* 2005). Although it is not already verified, it is speculated that ChsA contributes to motility interacting with the chemotactic machinery as was described in *P. aeruginosa*.

The bacterial chemotaxis provides a resource for responding to gradients of potential nutrients in the environment, by moving towards or away from these substances (Adler, 1966). The root exudates are an important source of nutrients for the microorganisms in the rhizosphere, and participate in the colonization process through chemotaxis of microorganisms in the soil (Campbell and Greaves, 1990; Hiroyuki *et al.*, 1998; Lynch and Whipps, 1990).

It is shown in this chapter that root exudates from different varieties of strawberry caused positive chemotaxis of different strains of *A. brasilense*. This effect was greater with endophytic strains than with those that colonize externally the root surface. Similar results were observed in the study of the chemostatic response of endophytic and rhizospheric bacteria toward root exudates in rice (Bacilio-Jiménez *et al.*, 2003). It was observed that the chemotactic effect of local strains of *A. brasilense* towards strawberry root exudates depended on the plant variety, where the best chemostactic response was obtained with exudates of the variety 'Camarosa.' However, differences in responses may be due to the origin of the bacterial strains (rhizospheric or endophytic). Furthermore, the specificity of strains probably reflects the adaptation of the bacterium to the nutritional conditions provided by the plant and can, thus, play an important role in the establishment of *Azospirillum* in the rhizosphere of the host, as previously indicated by Reinhold et al. (1985).

It was also verified that the concentration of root exudates affected chemotaxis response, as the best response was obtained when using a low concentration. These results agree with those previously reported in other studies regarding the ability of *Azospirillum* to detect potential sources of nutrients at lower concentrations than other soil microorganisms, favouring its survival in natural environments (Chet and Mitchell, 1976; Roszak and Colwell, 1987). Also, the production time of the exudates had effects on chemotaxis as a major chemotactic result was toward the exudates produced within the first two weeks of plant growth. This also coincides with the observations of Bacilio-Jimenez et al. (2003), working with root exudates of hydroponic rice.

In this case, the crop root exudates from strawberry in hydroponic conditions showed a decreased concentration of carbohydrates with the production time thereof, and an increase in protein concentration. This would indicate that the sugar residues have a greater attractive effect than protein, al least in the early stages of the partnership. These results partially coincide with those obtained by Bacilio-Jimenez et al. (2003) in rice hydroponics, who detected a higher concentration of sugars and amino acids in the exudates obtained during the first two weeks of cultivation. In other works it has been observed that during germination of wheat seeds, the concentration of carbohydrates was higher in the first seven days than in the second and fourth weeks of plant growth (Jones and Darrah, 1993; Prikryl and Vankura, 1980).

The decrease in total sugar concentration determined in hydroponic culture after the second week was probably caused by two potential mechanisms: the accumulation of high levels of organic substances in the vicinity of the roots that suppressed the release of additional organic compounds (Jones and Darrah, 1993; Prikryl and Vankura, 1980), and the re-absorption of organic compounds by the plant (Guckert *et al.*, 1991; Jones and Darrah, 1993). Moreover, the growing conditions conducted in this study could have influenced the amount and composition of the exudates released by the roots, as also indicated by Jones and Darrah (1993). Besides, the lack of enough oxygen in the tubes containing the plants may have contributed to the decrease in root exudation over the time.

However, the results showed in this chapter are consistent with those previously reported by Reinhold et al. (1985) for different organic compounds of root exudates of other crops; they indicate that the association of *Azospirillum* with the host plant is of type strain-specific.

By assessing the chemotactic activity of different strains of *A. brasilense* it was possible to observe the affinity of certain genotype in the association *Azospirillum*-strawberry. Considering that *A. brasilense*, posses biot-echnological application, addressing to a sustainable agriculture, determining the genes and mechanisms involved in chemotaxis response, as well as the activity of strains to root exudates may represent an initial step in selecting them for use as inoculants in different crops.

Acknowledgments

The laboratory work of B.E.B was supported by grant CONACYT Ref. 49227-Z. Experimental work developed in Argentina was partially supported by CIUNT and ANPCyT grants (PICT 2007 N°472). The authors acknowledge the support of CYTED through the DIMIAGRI network project.

References

Adler, J. (1966). Chemotaxis in bacteria. *Science*, *153*, 708-716.

Ahmad, F., Ahmad, I. & Khan, M. S. (2006). Screening of free-living rhizospheric bacteria for their multiple plant growth promoting activities. *Microbiol. Res.*, 36, 1-9.

Alexandre, G. (2007). Chemosensory behaviour in *Azospirillum brasilense. Azospirillun* VII and related PGPR. *Genomics, molecular ecology, plant responses and agronomic significance.* Rhizospphere II satellite international workshop. Montepellier, France.

Alexandre, G. & Zhulin I. B. (2001). More than one way to sense chemicals. *J. Bacteriol.*, *183*, 4681-4686.

Alexandre, G., Greer, S. E. & Zhulin I. B. (2000). Energy taxis is the dominant behaviour in *Azospirillum brasilense J. Bacteriol.*, *182*, 6042-6948.

Asghar, H. N., Zahir, Z. A., Arshad, M. & Khaliq A. (2002). Relationship between *in vitro* production of auxins by rhizobacteria and their growth promoting activities in *Brassica juncea* L. *Biol. Fertil. Soil*, *35*, 231-237.

Bacilio-Jimenéz, M., Aguilar Flores, S., Ventura- Zapata, Pérez-Campo S. & Bouquelet Zenteno, E. (2003). Chemical characterization of root exudates from rice (*Oryza sativa*) and their effects on the chemotactic response of endophytic bacteria. *Plant and Soil*, *249*, 271-277.

Bacilio-Jimenéz, M., Aguilar Flores, S., Ventura- Zapata, Pérez-Campo S. & Bouquelet Zenteno, E. (2001). Endophytic bacteria in rice seeds inhibit early colonization by *Azospirillum brasilense*. Soil. *Biol. Biochem.*, *33*, 167-172.

Barak, R., Nur, I., Okon, Y. & Henis Y. (1982). Aerotactic response of *Azospirillum brasilense. J. Bacteriol., 152*, 643-649.

Barbour, W. M., Hattermann, D. R. & y Stacey, G. (1991). Chemiotaxis of *Bradyrhizobium Japanicum* to soybean exudates. *Appl. Environ. Microbiol. 57*, 2635-2639.

Bashan Y. & Holguin, G. (1994). Root-root travel of the beneficial bacterium *Azospirillum brasilense. Appl. Environ. Microbiol., 60*, 2120-2131.

Bashan, Y. (1986). Enhancement of wheat roots colonization and plant development by *Azospirillum brasilense* Cd following temporary depression of the rhizosphere microflora. *Appl. Environ. Microbiol., 51*, 1067-1071.

Bashan, Y., Holguin, G. & de-Bashan, L. E. (2004). *Azospirillum*-plant relationships: physiological, molecular, agricultural, and environmental advances. *Can. J. Microbiol., 50*, 521-577.

Bible, A. N., Stephens, B. B., Ortega, D. R., Xie, Z. & Alexandre, G. (2008). Function of a chemotaxis-like signal transduction pathway in modulating motility, cell clumping and cell length in the alpha-proteobacterium *Azospirillum brasilense J. Bacteriol., 190*, 6365-6367.

Biswas, J. C., Ladha, L. K. & y Dazzo, F. B. (2000). Rhizobia inoculation improves nutrient uptake and growth of lowland rice. *J. Soil. Sci., 64*, 1644-1650.

Bobrov, A. G., Kirillina, O. & Perry, R. D. (2005). The phosphodiesterase activity of the HmsP EAL domain is required for negative regulation of biofilm formation in *Yersinia pestis. FEMS Microbiol. Lett., 247*, 123-130.

Bradford, M. M. (1976). A rapid and sensitive method for the quantification of microgram quantities of protein utilizing the principle of protein-dye binding. *Anal. Biochem., 72*, 248-264.

Burdman, S., Volpin, H., Kigel, J., Kapulnick, Y. & Okon, Y. (1994). Promotion of nod Genes inducers and inoculation in common bean (*Phaseolus vulgaris*) roots inculated with *Azospirillum brasilense* Cd. *Appl. Environ. Microbiol., 32*, 3030-3033.

Campbell, R. & y Graves, M. P. (1990). Anatomy and community structure of the rhizosphere. In the rhizosphere. *Ed. J Lynch.*, 11-34. John Wiley and Sons. Inc., New York.

Carreño-López, R., Sánchez, A., Camargo, N., Elmerich, C. & Baca, B. E. (2009). Characterization of *chsA*, a new gene controlling chemotactic response in *Azospirillum brasilense. Arch. Microbiol., 191*, 501-507.

Cattelan, A. J., Hartel, P. G. & y Fuhrmann, J. J. (1999). Screening for plant growth-Promoting rhizobacteria to promote early soybean growth. *Soil Sci. Soc. Am. J.*, *63*, 1670-1680.

Chet, I. & y Mitchell, R. (1976). Ecological aspects of microbial chemotactic behaviour. *Annu. Rev. Microbiol.*, *30*, 221-239.

Croes, C., Moens, S., Van Bastelaere, E., Vanderleyden, J. & Michiels, K. M. (1993). The polar flagellum mediates the adsorption to wheat roots. *J. Gen. Microbiol.*, *139*, 2261-2269.

Curl, E. A. & Truelove, B. (1986). *The rhizosphere*. pp228. Springer-Verlag. Berlin, Germany.

Davison, J. (1988). Plant beneficial bacteria. *Bio-Technology*, *6*, 182-286.

De Weert, S., Vermeiren, H., Mulders I. H. M., Kuiper, I., Hendrickx, N., Bloemberg, G. V., Vanderleyden, J., De Mot, R. & Lugtenberg, B. J. J. (2002). Flagella driven chemotaxis towards exudate components is an important trait for tomato root colonization by *Pseudomonas fluorescens*. *Mol. Plant-Microbe Interact.*, *15*, 1173-1180.

Dobbelaere, S., Croonenborghs, A., Thys, A., Ptacek, D., Vanderleyden, J., Dutto, P., Labendera-Gonzalez, C., Caballero-Mellado, J., Aguirre, F., Kapulnik, Y., Brener, S., Burdman, S., Kadouri, D., Sarig, S. & y Okon, Y. (2001). Response of agronomically important crops to inoculation with *Azospirillum*. *Aust. J. Plant. Physiol.*, *28*, 871-879.

Dobbelaere, S., Vanderleyden, J. & y Okon, Y. (2003). Plant growth effects of diazotrophs in the rhizosphere. *Crit. Rev. in Plant Sci.*, *22*, 107-149.

Döbereiner, J. & y Pedrosa, F. (1987). Nitrogen-fixing bacteria in non-leguminous crop plant. Brock/Spring. *Contemporary/Bioscience.*, *155*.

Egamberdiyeva, D. (2007). The effect of plant growth promoting bacteria on growth and Nutrient uptake of maize in two different soils. *Appl. Soil*, *36*, 184-189.

Galperin, M. Y. (2005). A census of membrane-bound and intracellular signal transduction proteins in bacteria: bacterial IQ, extroverts and introverts. *BMC Microbiol.*, *5*, 35-41.

Gravel, V., Antoun, H. & y Tweddell, R. J. (2007). Growth stimulation and fruit yield improvement of greenhouse tomato plants by inoculation with *Pseudomonas putida* or *Trichoderma atroviride*: Possible role of indoleacetic acid (IAA). *Soil. Biol. Biochem.*, *39*, 1968-1977.

Greer, S. E., Stephens, B. B. & Alexandre, G. (2004). An energy transducer promotes root colonization by *Azospirillum brasilense*. *J. Bacteriol.*, *186*, 6595-6604.

Greer-Phillips, S. E., Alexandre, G., Taylor, B. L. & Zhulin, I. B. (2003). Aer and Tsr guide *Escherichia coli* in spatial gradients of oxidizable substrates. *Microbiology, 149*, 2661-2667.

Guckert, A., Chavanon, M., Mench, M., Morel, J. L. & y Villenium, G. (1991). Root exudation in *Beta vulgaris*: A comparation with *Zea mays*. In *plant roots and their environment.* Michael, B.V.B.L., Person, H. (Eds). 449-445. Elsevier Science publishers, Amsterdam.

Guocheng, H. & Cooney, J. J. (1993). A modified capillary assay for chemotaxis. *J. Industrial Microbiol., 12*, 396-398.

Hall, P. G. & Krieg N. R. (1983). Swarming of *Azospirillum brasilense* on solid media. *Can. J. Microbiol., 29*, 1592-1594.

Hauwaerts, D., Alexandre, G., Das, S. K., Vanderleyden, J. & Zhulin I. B. (2000). A major chemotaxis gene cluster in *Azospirillum brasilense* and relationships between chemotaxis operons in α-proteobacterias. *FEMS Microbiol. Let., 208*, 61-67.

Heinrich D. & y Hess D. (1985). Chemotactic attraction of *Azospirillum lipoferum* by wheat roots and characterization of some attractants. *Can J Microbiol., 31*, 26-31.

Hickman, J. W., Tifrea, D. F. & Harwood, C. S. (2005). A chemosensory system that regulates biofilm formation through modulation of cyclic diguanylate levels. *Proc. Natl. Acad. Sci.*, USA *102*, 14422–27.

Hiroyuki, F., Masao, S., Hidenori, O., Yasufumi, U., Tadayoshi, S. & y Tatsuhiko, M. (1998). Chemotactic response to amino acids of fluor-escent pseudomonas isolated from Spinach roots grown in soils with different salinity levels. *Soil Sci. Plant Nutr., 44*, 1-7.

Hoagland, D. R. (1975). *Mineral nutrition*. In laboratory experiments in plant physiology. Eds. P. B. De Kaufman J., Labavich A., Anderson-Prouty, & Ghosheh. N. S. 129-134. Macmillan Publishing Co. Inc., New York.

Jeun, Y. C., Park, K. S., Kim, C. H., Fowler, W. D. & y Kloepper, J. W. (2004). Cytological observations of cucumber plants during induced resistance elicited by rhizobacteria. *Biol. Control., 29*, 34-42.

Jonas, K., Melefors, Ö. & Römling, U. (2009). Regulation of c-di.GMP metabolism in biofilms. *Future Microbiol., 4*, 1-18.

Jones, D. L. & Darrah, P. R. (1993). Resorption of organic compounds by roots of Zea mays L. and its consequences in the rhizosphere. *Plant Soil*, 153, 47-59.

Kato, J., Kim, H. E., Takiguchi, N., Kuroda, A. & Ohtake, H. (2008). *Pseudomonas aeruginosa* as a model microorganism for investigation of chemotactic behaviours in ecosystem. *J. Biosc and Bioeng., 106*, 1-7.

Khalid, A., Arshad, M. & y Zahir, Z. A. (2004). Screening plant growth-promoting rhizobacteria for improving growth and yield of wheat. *J. Appl. Microbiol., 96*, 473-480.

Kloepper, J. W., Lifshitz, R. & y Zablotowicz, R. M. (1989). Free-living bacterial inocula for enhancing crop productivity. *Trends Biotechnol., 7*, 39-43.

Kloepper, J. W. & Beauchamp, C. J. (1992). A review of issues related to measuring of plant roots by bacteria. *Can. J. Microbiol., 38*, 1219-1232.

Kozdroja, J., Trevorsb, J. T. & y van Elsasc, J. D. (2004). Influence of introduced potential biocontrol agents on maize seedling growth and bacterial community structure in the rhizosphere. *Soil. Biol. Biochem., 36*, 1775-1784.

Lucy, M., Reed, E. & y Bernard, R. Glick. (2004). Applications of free living growth-promoting rhizobacteria. *Antony Van Leeuwenhoek, 86*, 1-25.

Lynch, J. M. & y Whipps, J. M. (1990). Substrate flow in the rhizosphere. *Plant Soil, 129*, 1-10.

Mandimba, G., Heulin, T., Bally, R., Guckert, A. & Balandreau, J. (1986). Chemotaxis of free-living nitrogen-fixing bacteria towards maize mucilage. *Plant Soil, 90*, 129-139.

Milcamps, A., van Dommelen, A., Stigler, J., Vanderleyden, J. & de Bruijn, F. J. (1996). The *Azospirillum brasilense rpoN* gene is involved in nitrogen fixation nitrate assimilation ammonium uptake and flagellar biosynthesis. *Can. J. Microbiol., 42*, 3205-3218.

Mrkovacki, N. & y Milic, V. (2001). Use of *Azotobacter chroococcum* as potentially useful in agricultural application. *Ann. Microbiol., 51*, 145-158.

Niu, C., Graves, D. J., Mokuolu, F. O., Gilbert, S. E. & y Gilbert, E. S. (2005). Enhanced swarming of bacteria on agar plates containing the surfactant Tween 80. *J. Microbiol. Met., 62*, 129-132.

Okon, Y. & Vanderleyden, J. (1997). Root-associated *Azospirillum* species can stimulate plants. *ASM News, 63*, 366-370.

Okon, Y., Cakmakci, L., Nur, I. & Chet, I. (1980). Aerotaxis and chemiotaxis of *Azospirillum brasilense. Microbiol. Ecol., 6*, 277-280.

Ozturk, A., Caglar, O. & y Sahin, F. (2003). Yield response of wheat and barley to inoculation of plant growth promoting rhizobacteria at various levels of nitrogen fertilization. *J. Plant. Nutr. Soil Sci., 166*, 262-266.

Palleroni, J. N. (1976). Chamber for bacterial chemotaxis experiments. *Appl. Environ. Microbiol., 32*, 729-730.

Pedraza R. O., Motok, J., Salazar, S. M., Ragout, A. L., Mentel, M. I., Tortora, M. L., Guerrero-Molina, M. F., Winik, B. C. & Díaz-Ricci, J. C. (2009).

Growth-promotion of strawberry plants inoculated with *Azospirillum brasilense*. *World J. Microbiol. Biotechnol.*, *26*, 265-272.

Pedraza, R. O., Motok, J., Tortora, M. L., Salazar, S. L. & y Díaz Ricci, J. C. (2007). Natural occurrence of *Azospirillum brasilense* in strawberry plants. *Plant soil*, *295*, 169-178.

Prikryl, Z. & y Vankura, V. (1980). Roots exudates of plant. VI. Wheat root exudation as Dependent on growth. Concentration gradient of exudates and the presence of bacteria. *Plant Soil*, *57*, 1287-1290.

Raju, N. S., Niranjana, S. R., Janardhana, G. R., Prakash, H. S., Shetty, H. S. & y Mathur, S. B. (1999). Improvement of seed quality and field emergence of *Fusarium moniliforme* infected sorghum seeds using biological agents. *J. Sci. Food. Agric.*, *79*, 206.212.

Rashid, M. H. & y Kornberg, A. (2000). Inorganic polyphosphate is needed for swimming, swarming, and twitching motilities of *Pseudomonas aeruginosa*. PNAS, *97*, 4885-4890.

Reinhold, B., Hurek, T. & y Fendrik, I. (1985). Strain-Specific Chemotaxis of *Azospirillum* spp. *J. Bacteriol.*, *162*, 190-195.

Roszak, D. B. & y Colwell, R. R. (1987). Survival strategies of bacteria in natural environments. *Microbiol. Rev.*, *51*, 365-379.

Salantur, A., Ozturk, A. & y Akten, S. (2006). Growth and yield response of spring wheat (*Triticum aestivum* L.) to inoculation with rhizobacteria. *Plant Soil Environ.*, *52*, 111-118.

Schimer, T. & Jenal, U. (2009). Structural and mechanistic determinants of c-di-GMP signaling. *Nature Rev. Microbiol.*, *7*, 724-735.

Schmidt, A. J., Ryenkov, D. A. & Gomelsky, M. (2005). The ubiquitous protein domain EAL is a cyclic diguanylate-specific phosphodiesterase, enzymatically active and inactive EAL domains. *J Bacteriol.*, *187*, 4774-4781.

Shaharoona, B., Arshad, M., Zahir, Z. A. & y Khalid, A. (2006). Performance of *Pseudomonas* spp. containing ACC-deaminase for improving growth and yield of maize (*Zea mays* L.) in the presence of nitrogenous fertilizer. *Soil. Biol. Biochem.*, *38*, 2971-2975.

Siddiqui, I. A. & y Shaukat, S. S. (2002). Mixtures of plant disease suppressive bacteria enhance biological control of multiple tomato pathogens. *Biol. Fertil. Soil*, *36*, 260-268.

Simm, R., Morr, M., Kader, A., Nimtz, M. & Römling, U. (2004). GGDEF and EAL domains inversely regulate cyclic di-GMP levels and transition from sessility to motility. *Mol. Microbiol.*, *53*, 1123-1134.

Stephens, B. B., Loar, S. N. & Alexandre, G. (2006). Role of CheB and CheR in the complex chemotactic and aerotactic pathway of *Azospirillum brasilense*. *J. Bacteriol.*, *188*, 4759-4768.

Stock, J. B. & Surette, M. G. (1996). Chemotaxis, p. 1103-1129. In: R., Cutiss, III., J. L., Ingraham, E. C. C., Lin, K. B., Low, B., Magasanik, W. S., Rezinikoff, M., Riley, M. & Schaechter, H. E. Umbarger, (Eds.), *Escherichia coli* and *Salmonella* cellular and molecular biology, 2nd ed. American Societety for Microbiology, Whashington, D. C.

Szurmant, H. & Ordal, G. H. (2004). Diversity in chemotaxis mechanisms among the bacteria and archaea. *Microbiol. Mol. Biol. Rev.*, *68*, 301-319.

Taylor, B. & Zhulin, I. (1999). PAS domains: internal sensors of oxygen redox potential and light. *Microbiol. Mol. Biol. Rev.*, *63*, 479-506.

Taylor, B. L. & Zhulin, I. B. (1998). In search of higher energy: metabolism-dependent behaviour in bacteria. *Mol. Microbiol.*, *28*, 683-690.

Van Bastelaere, E., Lambrecht, M., Vermeiren, H., Van Dommlen, A., Keijers, V., Proost, P. & Vanderleyden, J. (1999). Characterization of a sugar-binding protein from *Azospirillum brasilense* mediated chemotaxis to and uptake of sugars. *Mol. Microbiol.*, *32*, 703-714.

Van Rhijn, P., Vanstockem, M., Vanderleyden, J. & De Mot, R. (1990). Isolation of behavioural mutants of *Azospirillum brasilense* by using *Tn5-lacZ*. *Appl. Environ. Microbiol.*, *4*, 990-996.

Vanbleu, E. & y Vanderleyden, J. (2003). Molecular Genetics of rhizosphere and plant root colonization. 1. Kluwer Academics Publishers. Netherlands.

Vande Broek, A., Lambrecht, M. & Vanderleyden, J. (1998). Bacterial chemotactic motility is important for the initiation of wheat root colonization by *Azospirillum brasilense*. *Microbiol.*, *144*, 2599-2606.

Vessey, K. (2003). Plant growth promoting rhizobacteria as biofertilizers. *Plant and Soil*, *255*, 571-586.

Vikram, A. (2007). Efficacy of phosphate solubilizing bacteria isolated from vertisols on growth and yield parameters of sorghum. *Res. J. Microbiol.*, *2*, 550-559.

Weller, D. M. & y Tamashow, L. S. (1994). Current challenges in introducing beneficial microorganisms into the rhizosphere. In: O., ´Gara, F., Dowling, D. N. & Boesten, B. (Eds). *Molecular ecology of rhizosphere microorganism. Biotechnology and the release of GMOs*. 1-18. Weinheim, Germany: VCH Verlags Gesellschaft mbH.

Zhulin, I. B, Bespalov, V. A., Johnson, M. S. & Taylor, B. L. (1996). Oxygen taxis and proton motive force in *Azospirillum brasilense*. *J. Bacteriol.*, *178*, 5199-5204.

Zhulin, I. B. & Armitage, J. P. (1992). The role of taxis in ecology of *Azospirillum*. *Symbiosis*, *13*, 199-206.

Zhulin, I. B. & Armitage, J. P. (1993). Motility, chemokinesis, and methylation-independent chemotaxis in *Azospirillum brasilense*. *J. Bacteriol.*, *175*, 952-958.

Reviewed by Dr. Katia Regina Dos Santos Teixeira. EMBRAPA Agrobiologia, Km 7 BR 465, Seropedica, Rio de Janeiro, CEP 23851-970, Brazil. E-mail, katia@cnpab.embrapa.br

In: Chemotaxis: Types, Clinical Significance… ISBN: 978-1-61728-495-3
Editor: T.C. Williams, pp.85-108

Chapter 3

CD46, Chemotaxis and MS– Are These Linked?

Siobhan Ni Choileain[1,2,3], Jillian Stephen[2], Belinda Weller[4] and Anne L. Astier[1,2,3*]

[1]Institute for Immunology and Infection Research
[2]MRC Centre for Inflammation Research
[3]Centre for MS Research, University of Edinburgh, Edinburgh, UK
[4]MS clinic, Division of Clinical Neuroscience, Western General Hospital, Edinburgh, UK

Abstract

Multiple Sclerosis (MS) is an autoimmune disease characterized by chronic inflammation of the brain. One of the main occurrences is the breach of the blood-brain-barrier, resulting in the entrance of inflammatory cells, which perpetuate the inflammation occurring in the brain. One of the mechanisms controlling cell migration is mediated by the release of small soluble molecules, called chemokines and by the expression of their corresponding receptors, the chemokine receptors. Hence, the conjoint expression of chemokine receptors and production of their relevant chemokines will direct the migration of cells towards the site of inflammation. Among the cells involved in the pathogenesis of

* Corresponding author: E-mail: a.astier@ed.ac.uk

MS, T cells have been shown to play an important role. Indeed, T cell activation is crucial for the immune homeostasis, notably through the balance of effectors T cells (Teff) and regulatory T cells (Tregs). In MS, defective Treg functions have been observed, which might partly explain the increased inflammation seen in MS. Among Tregs, Tr1 cells are characterized by the secretion of large amount of IL-10, an anti-inflammatory cytokine. The molecule CD46 is a regulator of complement activity. However, its activation also promotes T cell activation and differentiation toward Tr1 cells. This pathway is altered in MS, as the amount of IL-10 produced by CD46-activated T cells is largely reduced. This chapter will discuss preliminary evidence suggestive of a role of CD46 in the control of chemotaxis of activated T cells, which might play a role in MS pathogenesis.

Introduction

Worldwide, multiple sclerosis (MS) affects 2.5 million individuals. There is no cure, and as MS affects young adults in the productive years of their career, it has a significant social and economic impact. It is a complex autoimmune disease with chronic inflammation of the central nervous system (CNS) [1-4]. Migration of inflammatory cells to the brain is a critical step in the pathogenesis of MS, as cells must cross the BBB (6) in order to reach the CNS. Among the molecules involved in this process, integrins as well as chemokines and chemokine receptors are key molecules in orchestrating cellular migration [5].

We and others have shown that CD46 acts as a T cell co-stimulatory molecule [6-8]. It also induces differentiation toward a regulatory Tr1 phenotype characterized by the secretion of the anti-inflammatory cytokine IL-10 [9] and of Granzyme B [10]. CD46 also modulates cytokine production and chemokine secretion by dendritic cells (DCs) [11]. We have shown that CD46 functions are dysregulated in MS. First, IL-10 production by CD46-stimulated T cells is impaired [12-14]. Second, higher levels of the chemokines CCL3 and CCL5 and of the proinflammatory cytokine IL-23 are produced by CD46-activated DCs from patients with MS than DCs from healthy donors [11]. Thus, in MS, CD46 activation induces a proinflammatory phenotype in both T cells and DC. Furthermore, a recent report indicates that CD46 stimulation modulates integrin and chemokine receptor expression on T cells [15], and we also observed a modulation of chemokine receptor expression upon CD46 activation. This suggests that links exist between CD46 and chemotaxis, and this most likely contributes to the inflammation observed in MS. Understanding the factors that control T cell migration in MS is of critical importance to understanding the inflammation that occurs in this disease and might provide new means to develop therapeutics targets.

Multiple Sclerosis

Worldwide, MS affects 2.5 million individuals, with 85,000 patients in the UK and with Scotland having the highest global rate per capita. There is no cure, and new approaches desperately need to be developed for the treatment of this disease.

MS results from inflammation in the brain. It is a complex neurodegenerative disorder with autoimmune, genetic and environmental factors [4, 16-19]. Central nervous system (CNS) inflammation is notably driven by T cells, which recognize and target the myelin protein that surrounds and protects the nerve fibers in the brain. This leads to repeated inflammation of the nerves and the eventual slowing or stopping of the nerve impulses, causing the symptoms of MS.

To gain access to the brain, T cells must cross the blood-brain-barrier (BBB). The BBB is composed of tight junctions that restrict exchanges between circulating blood and the cerebrospinal fluid (CSF) in the CNS. In MS, there is a breach of the BBB, resulting in the passage of inflammatory cells into the CNS. The migration of inflammatory cells across the BBB and into the brain is therefore a key step in the pathogenesis of MS. Determining the factors that control T cell migration in MS is of critical importance to understanding the inflammation that occurs in this disease. Cell migration involves chemokines, chemokine receptors and integrins [20].

Chemokines

Chemokines are small proteins, with a molecular weight of around 8-10 kDa, with chemoattractant properties that stimulate the migration and activation of cells, a process known as chemotaxis [21]. Hence, cells secreting chemokines will attract cells expressing the corresponding chemokine receptors. Chemokines are produced by a number of different cells, including lymphocytes, and play an important role in the trafficking of cells to sites of inflammation. Chemokines generally possess four conserved cysteine residues in their amino acid sequence. These conserved cysteines allow the proper formation of their conformational shape that will confer their biological role. Chemokines can be split into 4 subfamilies, depending on the orientation of these conserved residues. They are known as the CC, CXC, CX_3C and C chemokines [22].

Functionally, chemokines are mainly sub-classified into two main categories: pro-inflammatory or homeostatic. Pro-inflammatory chemokines are induced during the immune response to attract cells to the site of inflammation. Homeostatic chemokines are normally expressed on cells to control their migration during natural biological processes under 'healthy' conditions, such as immune surveillance.

Chemokines exert their biological function by binding to chemokine receptors [23]. Indeed, chemokine receptors are differentially expressed on T cells depending on their polarization and activation state [24-27]. For example, CXCR4 is mainly expressed on naive cells and binds to CXCL12 (or SDF-1), a potent chemoattractant for lymphocytes. CXCR3 is expressed on memory cells and activated T cells, and binds to several chemokines (CXCL9, CXCL10 and CXCL11). Th1 lymphocytes express CCR5 and CXCR3, and Th2 lymphocytes, CCR3, CCR4, and CCR8 [25, 28]. CXCR3 and its ligands promote a Th1 response while inhibiting a Th2 response by blocking Th2 migration. Although the main function of chemokines is to control cell migration, recent data suggest that they can also exert control on T cell proliferation and cytokine production [29-31]. Notably, CXCR3 can suppress T cell activation [32], while CXCR4 acts a costimulatory molecule and promotes IL-10 secretion [33-35]. Chemokine receptors are seven membrane spanning G-protein coupled receptors (GPCR) [23], and their activation and function are regulated by G-protein coupled receptor kinases (GRKs) and β-arrestins, which function to desensitize and internalize chemokine receptors [36], as further discussed in this chapter.

The chemokine-chemokine receptor interaction is very complex. There is more than fifty chemokines identified that can interact with twenty receptors [37]. Chemokines are very promiscuous and can bind to several receptors, and in turn, receptors bind to different chemokines. The versatility of this system ensures a fine-tuned immune response.

Chemokines and Their Receptors and MS

It has become well established that chemokines and chemokine receptors are key molecules in directing inflammatory cells, including T cells, into the CNS. Moreover, IFN-β, used as a therapy for MS, has been shown to regulate cell migration, by modulating chemokine receptor expression [38-40]. Intriguingly, chemokines are also essential in the development and physiology of the nervous system [41].

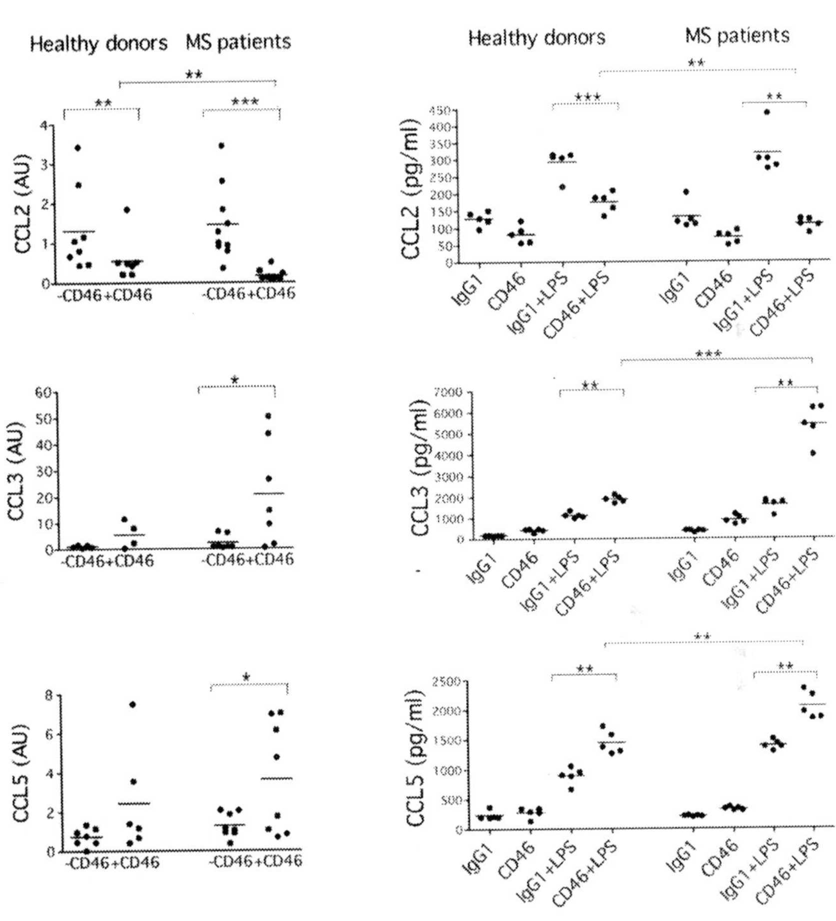

Figure 1. Chemokine production by CD46-activated mDCs. (A) The productions of CCL2, CCL3 and CCL5 by mature mDCs obtained by culture with LPS in presence or absence of CD46 antibodies were measured by qPCR (A) in a cohort of 8 healthy donors and 9 untreated patients with MS in the relapsing-remitting phase. Their relative expression compared to immature mDCs is plotted. (B) The amounts of CCL2, CCL3 and CCL5 secreted by mDC activated by CD46 (or irrelevant IgG1) were assessed by ELISA. *Reproduced with permission from Journal of NeuroImmunol, Vaknin-Dembinsky et al, 2008, 195:140-145.*

Abnormal levels of both chemokines and chemokine receptors have been identified in MS, as briefly summarized in the following paragraphs:

Chemokines: CCL3 (previously called MIP1-α) blockade with inhibitory antibodies inhibits the infiltration of mononuclear cells into the CNS and prevents the development of experimental autoimmune encephalomyelitis (EAE), a murine model of MS [42]. Indeed, increased expression of CCL3 and CCL5 (Rantes), two pro-inflammatory cytokines, were observed in the CSF of patients with an MS relapse compared to controls [43, 44]. We also observed that CD46-activated dendritic cells from MS patients secrete more CCL3 and CCL5 than cells from healthy controls, but less CCL2 (MCP-1) [11] (Figure 1). The role of the CCL2/CCR2 axis in MS is controversial. CCR2 also attracts inflammatory cells, such as T cells and monocytes. In EAE, an increased expression of CCR2 is observed after the initial attack [45]. Moreover, CCL2-deficient mice exhibit a delayed onset of active EAE and have reduced clinical signs [46]. However, there is a significant decrease in CCL2 in CSF from patients compared to controls [47]. Furthermore, CCL2 levels are higher in PBMC from untreated patients in stable phase, and are modulated by IFNβ or methylprednisolone treatments [48, 49]. These data might suggest a negative correlation between the levels of CCL2 and active MS. We showed that CCL2 levels are markedly reduced in patients with MS upon CD46 stimulation, again suggestive of a role of CD46 in MS pathogenesis [11]. CCL17, which binds to CCR4 on Treg cells, is over-expressed in MS [50]. Both CXCL12 (SDF-1) and its receptor (CXCR4) are involved in controlling T cell migration across the BBB. CXCL12 is highly expressed in MS lesions and present in the CSF of patients [51].

Chemokine receptors: CCR5 and CXCR3 expressions are increased on peripheral T cells from progressive MS patients. They are also observed in MS lesions, as are their ligands, CCL3 and CXCL10 (IP10) [52]. Expression of CCR5 has been shown to control sensitivity to apoptosis, and CCR5+ T cells from patients with primary progressive MS (PPMS) are more resistant to apoptosis than cells from healthy donors. This resistance to apoptosis could lead to their chronic persistence in peripheral blood [53]. Increased CXCR3 expression in peripheral blood $CD4^+$ T cells is associated with MS relapses [54]. Interestingly, an inverse correlation between Vitamin D levels, which has been linked to decreased MS risks, and CXCR3+ T cells was observed [55]. Our data suggest that CD46-activated T cells from MS patients in the relapsing-remitting stage express more CXCR3 than controls (Figure 2). In addition, CCR6 is known to direct inflammatory Th17 T cells into the CNS in an EAE model [56]. Chemokine receptors on myeloid cells also play a role in MS pathogenesis. For example, a role of chemokine-like receptor-1 (CMKLR1), expressed by myeloid DCs, in EAE has been demonstrated. CMKLR1 knockout mice develop less severe EAE than their wild-type counterparts [57]. Increased CCR8 expression on microglia and phagocytic macrophages has been detected in actively demyelinating MS lesions [58].

Overall, the balance of chemokines produced and of specific receptors expressed will control cell migration towards the site of inflammation and might be used to design new treatments for MS, as reviewed in [20].

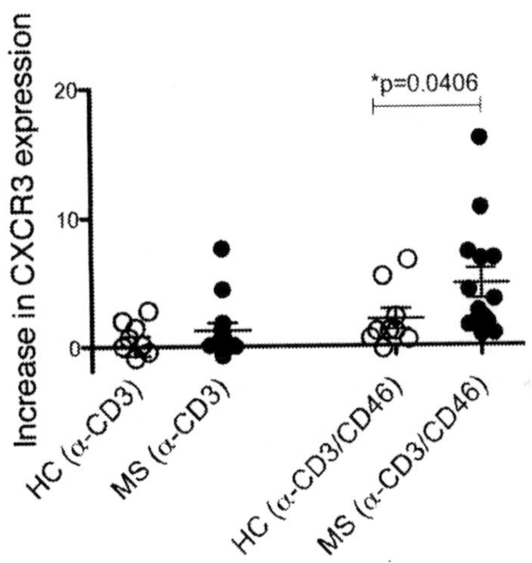

Figure 2. Increased CXCR3 expression by CD46-activated T cells in MS. CD4+ T cells were purified from a cohort of healthy controls (n=12) and patients with MS in the relapsing-remitting stage (n=9). Ethical approval was previously obtained from the Lothian NHS Ethics Board III. T cells were activated by pre-coated anti-CD3 antibodies (α-CD3), or anti-CD3/anti-CD46 antibodies in presence of IL-2 (α-CD3/CD46) to induce Tr1 cells as previously described [12]. The expression of CXCR3 was determined 2 days later by flow cytometry using anti-CXCR3-FITC antibodies (BD). The normalized MFI was calculated with the value obtained with isotype control antibodies - NormMFI= MFI(stained) – MFI(control). The increased expression compared to non-activated cells was calculated and plotted. Data were analyzed using the Mann-Whitney Mann-Whitney *U* test, a non-parametric test that does not assume Gaussian distribution.

GRKs and inflammation

As previously discussed, chemokine receptors are seven-membrane spanning G-protein coupled receptors (GPCR). They consist of an N-terminal portion, followed by seven helical transmembrane domains, with three extracellular and intracellular loops, and a C-terminal intracytoplasmic tail. The cytoplasmic tail contains several serine-threonine residues that can be phosphorylated. Indeed, signaling through GPCRs, coupled to heterotrimeric G_i proteins, is a crucial feature of the plasticity of the immune system and is involved in the regulation of inflammation (reviewed in [59]). Among the GPCRs, chemokine receptors have been the most studied in immune cells [21, 23, 60-62]. Once bound to their agonist ligands, GPCRs are phosphorylated by the serine/threonine kinases GRKs (G-protein coupled receptor kinases) [36, 63], resulting in their binding to β-arrestins and subsequent signaling impairment and internalization, a process known as desensitization [64, 65]. There are 7 types of GRKs referred to as GRK1-7, each with different expression profiles. Among them, GRK2, 3, 5 and 6 are ubiquitously expressed, but are expressed at particularly high levels in immune cells, and have been shown to regulate inflammation [59].

In addition GRK expression levels are tightly controlled in immune cells. T cell activation by PHA and anti-CD3 antibodies increases GRK expression and activity [66, 67]. In EAE, the decreased GRK2 expression leads to an earlier onset of the disease, which is correlated with an earlier increased T cell infiltration in the CNS [68]. Furthermore, mice heterozygous for deletion of the GRK2 gene (50% decrease) results in an increased chemotactic response to CCL3, CCL4 and CCL5, which signal via CCR5 [69]. Indeed, inhibiting GRK2/3 abrogates CCR5 phosphorylation [70] while GRK3 controls CXCR4 signalling [71]. However, GRK2-/+ mice do not develop relapses while wild-type animals do [68], suggesting a different and complex role of GRK2 depending on the course of the disease. Importantly, decreased levels of GRK2 and GRK6 were observed in T cells from patients with rheumatoid arthritis [72] and MS [68, 73]. T cells from these patients exhibit ~50% decreased GRK2 expression, correlated with an increased T cell chemotaxis to CCL3, CCL4, and CCL5 [68, 72]. Moreover, IFNβ used in the treatment of MS, modulates GRK2 and GRK3 [74]. *In vitro*, proinflammatory cytokines and oxygen radicals can decrease GRK2 levels [75]. Hence, the levels of GRKs seem to be crucial for the development of MS.

CD46

CD46 is a ubiquitously expressed transmembrane protein, first identified as a regulator of complement activation [76]. This type I membrane protein has cofactor activity for inactivation of C3b and C4b, as it allows their cleavage by protease I, hence protecting the cells from autologous killing. In addition, CD46 has been tagged as a 'pathogen's magnet' [77], as it acts as a receptor for many pathogens, including the Edmonston strain of measles virus, human herpesvirus-6, adenoviruses A and B, type IV pili of *Neisseria gonorrhoeae* and *Neisseria meningitidis* as well as group A streptococcus [77, 78]. The basic structure of CD46 is composed of four "short consensus repeats" (also called CCP or sushi domains) and a region rich in serine, threonine and proline (STP region) followed by a transmembrane segment and a short cytoplasmic tail. Multiple isoforms are produced due to the alternative splicing of various exons. In particular, two distinct intracytoplasmic tails (Cyt-1 and Cyt-2) co-exist [79]. These two isoforms are usually co-expressed in any given tissue, except for brain and kidney where a preferential expression of Cyt-2 is observed [80].

CD46 and T cells

Efficient T cell activation occurs upon TCR engagement and a concomitant stimulation with a costimulatory molecule, such as CD28 [81, 82]. We initially identified a novel role for CD46 in the regulation of the adaptive immune response by demonstrating its function as an additional costimulatory molecule for human T cells [6-8]. The molecule Crry, the murine analogue of CD46, also promotes T cell stimulation for murine T cells [83-85], highlighting a new biological function of these complement regulatory proteins (reviewed in [86-88]). Furthermore, CD46-costimulated human T cells in the presence of IL-2 acquire a T regulatory (Tr1) phenotype, secreting high levels of IL-10 [9] and granzyme B [10], although different conditions of stimulation can lead towards a Th1 phenotype [89]. CD46 costimulation also induces specific morphological changes of primary human T cells that spread *in vitro* [7], and alters T cell polarization [90]. We showed that the two intracellular tails of CD46 produced by alternative splicing, Cyt1 and Cyt2, have opposite effects on T cell-induced inflammation in an *in vivo* CD46-transgenic mouse model (mice only express CD46 in the testis) [8]. Each isoform differently controls the amounts of IL-2 and IL-10 secreted, and only the Cyt1 isoform induced T cell spreading.

CD46 and signal transduction

Signal transduction events mediated by CD46 is reviewed in [91]. Binding of pathogens to CD46 can trigger a calcium increase in epithelial cells [92], or recruit SHP-1 phosphatase to CD46 in human macrophages [93]. CD46 activation on antigen-presenting cells can either enhance or decrease IL-12/IL-23 production depending on the cell types and stimuli [11, 94-96]. In human T cells, several transducing molecules involved in TCR signaling are activated [6, 7, 89]. Both cytoplasmic tails associate with kinases [97]. The Cyt1 isoform binds to Dlg4, a scaffold signaling protein important for neuronal signaling and for the polarized expression of CD46 in epithelial cells [98]. Expression of Cyt1 in mouse macrophages enhances the production of nitric oxide in presence of IFNγ [99]. The Cyt2 isoform can be phosphorylated in both Jurkat and epithelial cells [92, 100]. A recent report also demonstrates an intriguing role of CD46 in regulating autophagy [101]. The authors show nicely that CD46-Cyt1 is interacting with the scaffold protein GOPC, linked to the autophagosome. In this system, autophagy was induced by pathogen binding to CD46. Together, these results argue in favor of active transduction pathways mediated by the two cytoplasmic isoforms. However, the precise transduction cascades initiated by the ligation of each isoform are still unknown.

MS and CD46

There are several links between CD46 and MS. CD46 is highly expressed at the BBB [102] and mediates access to the meninges as shown by transgenic mice expressing human CD46 (mice do not express CD46 except for in testis) which are then susceptible to meningococcal disease [103]. Of note, astrocytoma cells secrete CCL3 and CCL5 after CD46 engagement by measles virus or HHV-6, potentially involved in MS pathogenesis [104, 105]. Elevated levels of soluble CD46 are present in the serum and CSF of MS patients [106, 107]. Although controversial [108], associations between infection by HHV6 (that binds to CD46) and active MS have been described [107]. We demonstrated a defect in Tr1 differentiation in T cells from MS patients, characterized by a reduced IL-10 secretion [12, 109]. This has been recently confirmed by another study [14] and in a monkey model of MS [110]. This defect was associated with an increased RNA expression of the Cyt2 isoform [12], suggesting that the relative expression of the cytoplasmic isoforms might dictate T cell responses. We have recently confirmed the increased Cyt2 expression in peripheral T cells from patients at the protein level. Moreover, in MS, CD46-activated dendritic cells secrete high amounts of proinflammatory

IL-23, and of CCL3 and CCL5 chemokines [11]. Hence, CD46 appears to be crucial in the pathogenesis of MS, by switching the cells to a proinflammatory phenotype.

CD46 and chemotaxis

In primary human T cells, morphological changes are observed in CD46-costimulated cells, along with GTPase activation, which is involved in actin reorganization [7]. Our time-lapse experiments show an enhanced spreading of CD46-costimulated T cells with extremely dynamic adhesion and membrane extensions, suggestive of migratory capacities (unpublished data). In addition, CD46 activation of dendritic cells results in the secretion of chemokines and larger amounts of these chemokines are secreted in MS compared to healthy donors (Figure 1). CD46 stimulation upregulates α4β7 and CCR9 expression on T cells in presence of retinoid acid, suggestive of a gut-homing phenotype [15]. Interestingly, T cells isolated from the CSF also express α4β7 and CCR9 [111]. Our new data suggest that an increased expression of CXCR3 is observed in CD46-activated T cells from MS patients (Figure 2).

Overall, CD46 activation regulates chemokine secretion by dendritic cells, and chemokine receptor expression on activated T cells. Moreover, exacerbated chemokine secretion by DCs and increased CXCR3 expression on T cells were observed in MS. This suggests that CD46-induced chemotaxis likely plays an important role in mediating the inflammatory response in MS.

Integrins

Integrins are also involved in regulating cell migration. These receptors are composed of two chains, α and β, that are expressed at the cell surface. Their activation mediates attachment to proteins from the extracellular matrix (ECM) or to counter receptors expressed on other cells. This is a critical step for cell migration. In MS, both α4β1 and α4β7 integrins have been involved in migration in the CNS. The α4 chain of these integrins is also called VLA-4 (very late activation antigen 4). VLA4 recognizes its ligand VCAM-1 (vascular cell adhesion molecule-1) expressed on endothelial cells. The interactions between α4β1 or α4β7 and VCAM-1 mediate adhesion and migration of immune cells into the brain in EAE models [112-114]. As mentioned earlier, CD46-activation induces α4β7 expression on T cells [15], again suggestive of a role of CD46 in controlling T cell migration to the brain.

A

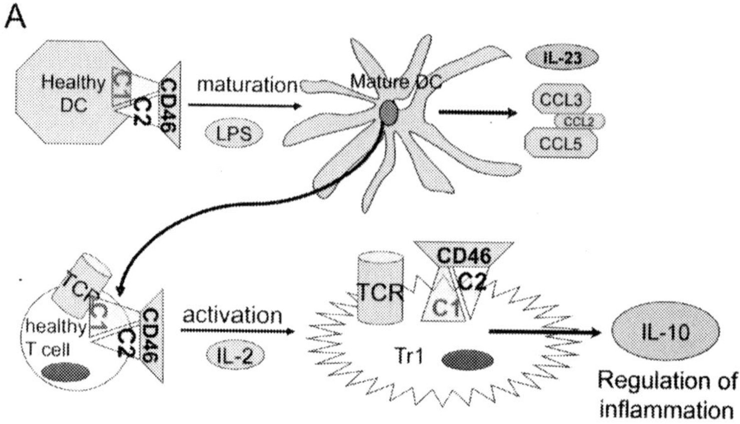

B

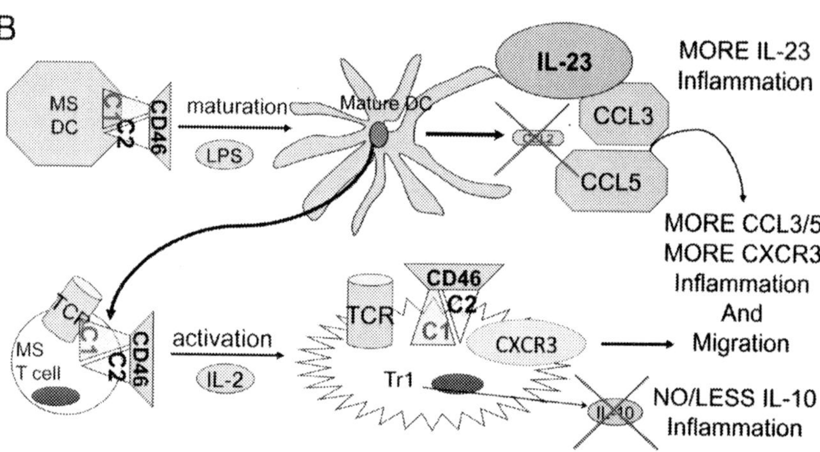

Figure 3. Altered CD46 pathways in MS. In healthy donors (A), CD46-activated dendritic cells (DCs) secrete small amounts of CCL3, CCL5 and IL-23. In contrast, CCL2 secretion by DCs is decreased when CD46 is engaged [11] and Figure 1. In MS (B), the secretion of CCL3, CCL5 and IL-23 by CD46-activated DCs is significantly increased compared to healthy donors, and production of CCL2 is abrogated [11], and Figure 1. Moreover, in MS, CD46-activated T cells express the pro-inflammatory chemokine receptor CXCR3 (see Figure 2). In healthy donors, CD46 activation of T cells leads to Tr1 differentiation and IL-10 production. In MS, IL-10 secretion upon CD46 is impaired [12-14, 110]. Hence, CD46 promotes a pro-inflammatory phenotype in MS, with upregulation of pro-inflammatory cytokines and chemokines/chemokine receptors. The further determination of the mechanisms controlling T cell activation and migration will help to understand the inflammation occurring in MS.

The identification of proteins involved in cell migration in MS has indeed provided novel therapeutic targets in the treatment of this disease. An example of one such treatment is the use of Natalizumab, which is a humanized monoclonal antibody specific for the α4 chain of the α4β1 and α4β7 integrins [115]. As a selective blocker of adhesion molecules, Natalizumab prevents the migration of T cells across biological barriers and suppresses T cell mediated immune responses. Moreover, treatment with Natalizumab also affects the profile of chemokines and cytokines produced [116].

Together these data suggest that CD46-induced integrins, chemokine and chemokine receptor expression may play an important role in T cell migration into the brain. Our current model on the role of CD46 in MS is summarized in Figure 3. Further understanding the inflammatory mechanisms observed in patients with MS could help to identify new therapeutic targets to treat this disease.

Conclusion

It is now well established that CD46 pathways are altered in MS. Both T cells and DCs play an important role in MS, with CD46 activation of these cells resulting in a pro-inflammatory response. It is likely that other immune cell types are affected, but further investigations need to be performed to address this issue. As discussed in this chapter, cell migration across the BBB is a crucial element in MS pathogenesis. It now emerges that CD46 has the ability to regulate the adaptive immune response, to fine-tune T cell responses but also to regulate chemokine secretion and chemokine receptor expression at the cell surface of activated DC and T cells. Moreover, the exacerbated expression of proinflammatory chemokines and chemokine receptors induced by CD46 in MS highlights its likely role in controlling chemotaxis. Further investigation of CD46-induced chemotaxis in MS will provide a better understanding of the inflammation that occurs during this disease. Identification of key chemokines and their receptors involved in mediating T cell migration in MS might also provide important therapeutic targets to help treat this disease.

Acknowledgment

These studies were supported by a research grant to ALA from the Multiple Sclerosis Society (859/07). We are very grateful to Nicola McLeod and Matthew Justin for their help in obtaining blood samples from patients with MS. We appreciate the help from Fiona Rossi and Shonna Johnston for cytometry analyses. We are very grateful to Dr. C. Rabourdin-Combe for the kind gift of the anti-CD46 monoclonal antibodies.

References

[1] Feldmann, M; Steinman, L. Design of effective immunotherapy for human autoimmunity. *Nature,* 2005, 435, 612-619.

[2] Hafler, DA; Slavik, JM; Anderson, DE; O'Connor, KC; De Jager, P; Baecher-Allan, C. Multiple sclerosis. *Immunol Rev,* 2005, 204, 208-231.

[3] Hohlfeld, R; Wekerle, H. Autoimmune concepts of multiple sclerosis as a basis for selective immunotherapy: from pipe dreams to (therapeutic) pipelines. *Proc Natl Acad Sci,* U S A, 2004, 101 Suppl 2, 14599-14606.

[4] Miller, D. Multiple sclerosis: new insights and therapeutic progress. *Lancet Neurol,* 2007, 6, 5-6.

[5] Yoshie, O; Imai, T; Nomiyama, H. Chemokines in immunity. *Adv Immunol,* 2001, 78, 57-110.

[6] Astier, A; Trescol-Biemont, M. C; Azocar, O; Lamouille, B; Rabourdin-Combe, C; Cutting edge: CD46, a new costimulatory molecule for T cells, that induces p120CBL and LAT phosphorylation. *J Immunol,* 2000, 164, 6091-6095.

[7] Zaffran, Y; Destaing, O; Roux, A; Ory, S; Nheu, T; Jurdic, P; Rabourdin-Combe, C; Astier, A. L; CD46/CD3 costimulation induces morphological changes of human T cells and activation of Vav, Rac, and extracellular signal-regulated kinase mitogen-activated protein kinase. *J Immunol,* 2001, 167, 6780-6785.

[8] Marie, JC; Astier, AL; Rivailler, P; Rabourdin-Combe, C; Wild, TF; Horvat, B. Linking innate and acquired immunity: divergent role of CD46 cytoplasmic domains in T cell induced inflammation. *Nat Immunol,* 2002, 3, 659-666.

[9] Kemper, C; Chan, AC; Green, JM; Brett, KA; Murphy, KM; Atkinson, JP. Activation of human CD4(+) cells with CD3 and CD46 induces a T-regulatory cell 1 phenotype. *Nature,* 2003, 421, 388-392.

[10] Grossman, WJ; Verbsky, JW; Tollefsen, BL; Kemper, C; Atkinson, JP; Ley, TJ; Differential expression of granzymes A and B in human cytotoxic lymphocyte subsets and T regulatory cells. *Blood,* 2004, 104, 2840-2848.

[11] Vaknin-Dembinsky, A; Murugaiyan, G; Hafler, DA; Astier, AL; Weiner, HL. Increased IL-23 secretion and altered chemokine production by dendritic cells upon CD46 activation in patients with multiple sclerosis. *J Neuroimmunol,* 2008, 195, 140-145.

[12] Astier, AL; Meiffren, G; Freeman, S; Hafler, DA; Alterations in CD46-mediated Tr1 regulatory T cells in patients with multiple sclerosis. *J Clin Invest,* 2006, 116, 3252-3257.

[13] Astier, AL. T-cell regulation by CD46 and its relevance in multiple sclerosis. *Immunology,* 2008, 124, 149-154.

[14] Martinez-Forero, I; Garcia-Munoz, R; Martinez-Pasamar, S; Inoges, S; Lopez-Diaz de Cerio, A; Palacios, R; Sepulcre, J; Moreno, B; Gonzalez, Z; Fernandez-Diez, B; Melero, I; Bendandi, M; Villoslada, P. IL-10 suppressor activity and ex vivo Tr1 cell function are impaired in multiple sclerosis. *Eur J Immunol,* 2008, 38, 576-586.

[15] Alford, SK; Longmore, GD; Stenson, WF; Kemper, C. CD46-induced immunomodulatory CD4+ T cells express the adhesion molecule and chemokine receptor pattern of intestinal T cells. *J Immunol,* 2008. 181, 2544-2555.

[16] Adorini, L. Immunotherapeutic approaches in multiple sclerosis. *J Neurol Sci,* 2004, 223, 13-24.

[17] McQualter, JL; Bernard, CC; Multiple sclerosis: a battle between destruction and repair. *J Neurochem,* 2007, 100, 295-306.

[18] Hafler, DA; Compston, A; Sawcer, S; Lander, ES; Daly, MJ; De Jager, PL; de Bakker, PI; Gabriel, SB; Mirel, DB; Ivinson, AJ; Pericak-Vance, MA; Gregory, SG; Rioux, JD; McCauley, JL; Haines, JL; Barcellos, LF; Cree, B; Oksenberg, JR; Hauser, SL. Risk alleles for multiple sclerosis identified by a genomewide study. *N Engl J Med,* 2007, 357, 851-862.

[19] Ramagopalan, SV; Maugeri, NJ; Handunnetthi, L; Lincoln, MR; Orton, SM; Dyment, DA; Deluca, GC; Herrera, BM; Chao, MJ; Sadovnick, AD; Ebers, GC; Knight, JC. Expression of the multiple sclerosis-associated MHC class II Allele HLA-DRB1*1501 is regulated by vitamin D. *PLoS Genet,* 2009, 5, e1000369.

[20] Hamann, I; Zipp, F; Infante-Duarte, C. Therapeutic targeting of chemokine signaling in Multiple Sclerosis. *J Neurol Sci,* 2008, 274, 31-38.

[21] Zlotnik, A; Yoshie, O; Chemokines: a new classification system and their role in immunity. *Immunity,* 2000, 12, 121-127.

[22] Luster, AD. Chemokines-chemotactic cytokines that mediate inflammation. *N Engl J Med,* 1998 338, 436-445.

[23] Murdoch, C; Finn, A. Chemokine receptors and their role in inflammation and infectious diseases. *Blood,* 2000, 95, 3032-3043.

[24] Moser, B. T-cell memory, the importance of chemokine-mediated cell attraction. *Curr Biol,* 2006, 16, R504-507.

[25] Sallusto, F; Lanzavecchia, A; Mackay, CR. Chemokines and chemokine receptors in T-cell priming and Th1/Th2-mediated responses. *Immunol Today,* 1998, 19, 568-574.

[26] Sallusto, F; Kremmer, E; Palermo, B; Hoy, A; Ponath, P; Qin, S; Forster, R; Lipp, M; Lanzavecchia, A. Switch in chemokine receptor expression upon TCR stimulation reveals novel homing potential for recently activated T cells. *Eur J Immunol,* 1999, 29, 2037-2045.

[27] Ebert, LM; McColl, SR. Coregulation of CXC chemokine receptor and CD4 expression on T lymphocytes during allogeneic activation. *J Immunol,* 2001, 166, 4870-4878.

[28] Culley, FJ; Pennycook, AM; Tregoning, JS; Hussell, T; Openshaw, PJ. Differential chemokine expression following respiratory virus infection reflects Th1- or Th2-biased immunopathology. *J Virol,* 2006, 80, 4521-4527.

[29] Contento, RL; Molon, B; Boularan, C; Pozzan, T; Manes, S; Marullo, S; Viola, A. CXCR4-CCR5: a couple modulating T cell functions. *Proc Natl Acad Sci U S A,* 2008, 105, 10101-10106.

[30] Whiting, D; Hsieh, G; Yun, JJ; Banerji, A; Yao, W; Fishbein, MC; Belperio, J; Strieter, RM; Bonavida, B; Ardehali, A. Chemokine monokine induced by IFN-gamma/CXC chemokine ligand 9 stimulates T lymphocyte proliferation and effector cytokine production. *J Immunol,* 2004, 172, 7417-7424.

[31] Marelli-Berg, FM; Okkenhaug, K; Mirenda, V. A two-signal model for T cell trafficking. *Trends Immunol,* 2007, 28, 267-273.

[32] Bromley, SK; Peterson, DA; Gunn, MD; Dustin, ML. Cutting edge: hierarchy of chemokine receptor and TCR signals regulating T cell migration and proliferation. *J Immunol,* 2000, 165, 15-19.

[33] Nanki, T; Lipsky, PE. Cutting edge: stromal cell-derived factor-1 is a costimulator for CD4+ T cell activation. *J Immunol,* 2000, 164, 5010-5014.

[34] Kumar, A; Humphreys, TD; Kremer, KN; Bramati, PS; Bradfield, L; Edgar, CE; Hedin, KE. CXCR4 physically associates with the T cell receptor to signal in T cells. *Immunity,* 2006, 25, 213-224.

[35] Kremer, KN; Kumar, A; Hedin, KE. Haplotype-independent costimulation of IL-10 secretion by SDF-1/CXCL12 proceeds via AP-1 binding to the human IL-10 promoter. *J Immunol,* 2007, 178, 1581-1588.

[36] Moore, CA; Milano, SK; Benovic, JL. Regulation of receptor trafficking by GRKs and arrestins. *Annu Rev Physiol,* 2007, 69, 451-482.

[37] Bromley, SK; Mempel, TR; Luster, AD. Orchestrating the orchestrators, chemokines in control of T cell traffic. *Nat Immunol,* 2008, 9, 970-980,

[38] Sorensen, TL; Roed, H; Sellebjerg, F. Chemokine receptor expression on B cells and effect of interferon-beta in multiple sclerosis. *J Neuroimmunol,* 2002, 122, 125-131.

[39] Sorensen, TL; Sellebjerg, F. Selective suppression of chemokine receptor CXCR3 expression by interferon-beta1a in multiple sclerosis. *Mult Scler,* 2002, 8, 104-107.

[40] Sellebjerg, F; Krakauer, M; Hesse, D; Ryder, LP; Alsing, I; Jensen, PE; Koch-Henriksen, N; Svejgaard, A; Soelberg Sorensen, P; Identification of new sensitive biomarkers for the in vivo response to interferon-beta treatment in multiple sclerosis using DNA-array evaluation. *Eur J Neurol,* 2009, 16, 1291-1298.

[41] Ransohoff, RM. Chemokines and chemokine receptors: standing at the crossroads of immunobiology and neurobiology. *Immunity,* 2009, 31, 711-721.

[42] Karpus, WJ; Lukacs, NW; McRae, BL; Strieter, RM; Kunkel, SL; Miller, SD. An important role for the chemokine macrophage inflammatory protein-1 alpha in the pathogenesis of the T cell-mediated autoimmune disease, experimental autoimmune encephalomyelitis. *J Immunol,* 1995, 155, 5003-5010.

[43] Bartosik-Psujek, H; Stelmasiak, Z. The levels of chemokines CXCL8, CCL2 and CCL5 in multiple sclerosis patients are linked to the activity of the disease. *Eur J Neurol,* 2005, 12, 49-54.

[44] Miyagishi, R; Kikuchi, S; Fukazawa, T; Tashiro, K. Macrophage inflammatory protein-1 alpha in the cerebrospinal fluid of patients with

multiple sclerosis and other inflammatory neurological diseases. *J Neurol Sci,* 1995, 129, 223-227.

[45] Glabinski, AR; Bielecki, B; O'Bryant, S; Selmaj, K; Ransohoff, RM. Experimental autoimmune encephalomyelitis, CC chemokine receptor expression by trafficking cells. *J Autoimmun,* 2002, 19, 175-181.

[46] Huang, DR; Wang, J; Kivisakk, P; Rollins, BJ; Ransohoff, RM. Absence of monocyte chemoattractant protein 1 in mice leads to decreased local macrophage recruitment and antigen-specific T helper cell type 1 immune response in experimental autoimmune encephalomyelitis. *J Exp Med,* 2001, 193, 713-726.

[47] Sorensen, TL; Ransohoff, RM; Strieter, RM; Sellebjerg, F. Chemokine CCL2 and chemokine receptor CCR2 in early active multiple sclerosis. *Eur J Neurol* 2004, 11, 445-449.

[48] Iarlori, C; Reale, M; De Luca, G; Di Iorio, A; Feliciani, C; Tulli, A; Conti, P; Gambi, D; Lugaresi, A. Interferon beta-1b modulates MCP-1 expression and production in relapsing-remitting multiple sclerosis. *J Neuroimmunol,* 2002, 123, 170-179.

[49] Moreira, MA; Tilbery, CP; Monteiro, LP; Teixeira, MM; Teixeira, AL. Effect of the treatment with methylprednisolone on the cerebrospinal fluid and serum levels of CCL2 and CXCL10 chemokines in patients with active multiple sclerosis. *Acta Neurol Scand,* 2006, 114, 109-113.

[50] Narikawa, K; Misu, T; Fujihara, K; Nakashima, I; Sato, S; Itoyama, Y. CSF chemokine levels in relapsing neuromyelitis optica and multiple sclerosis. *J Neuroimmunol,* 2004, 149, 182-186.

[51] Calderon, TM; Eugenin, EA; Lopez, L; Kumar, SS; Hesselgesser, J; Raine, CS; Berman, JW. A role for CXCL12 (SDF-1alpha) in the pathogenesis of multiple sclerosis, regulation of CXCL12 expression in astrocytes by soluble myelin basic protein. *J Neuroimmunol,* 2006, 177, 27-39.

[52] Balashov, KE; Rottman, JB; Weiner, HL; Hancock, WW. CCR5(+) and CXCR3(+) T cells are increased in multiple sclerosis and their ligands MIP-1alpha and IP-10 are expressed in demyelinating brain lesions. *Proc Natl Acad Sci U S A,* 1999, 96, 6873-6878.

[53] Julia, E; Edo, MC; Horga, A; Montalban, X; Comabella, M. Differential susceptibility to apoptosis of CD4+T cells expressing CCR5 and CXCR3 in patients with MS. *Clin Immunol,* 2009, 133, 364-374.

[54] Mahad, DJ; Lawry, J; Howell, SJ; Woodroofe, MN. Longitudinal study of chemokine receptor expression on peripheral lymphocytes in multiple

sclerosis, CXCR3 upregulation is associated with relapse. *Mult Scler,* 2003, 9, 189-198.

[55] Royal, W; 3rd, Mia, Y; Li, H; Naunton, K; Peripheral blood regulatory, T. cell measurements correlate with serum vitamin D levels in patients with multiple sclerosis. *J Neuroimmunol,* 2009, 213, 135-141.

[56] Reboldi, A; Coisne, C; Baumjohann, D; Benvenuto, F; Bottinelli, D; Lira, S; Uccelli, A; Lanzavecchia, A; Engelhardt, B; Sallusto, F. C-C chemokine receptor 6-regulated entry of TH-17 cells into the CNS through the choroid plexus is required for the initiation of EAE. *Nat Immunol,* 2009, 10, 514-523.

[57] Graham, KL; Zabel, BA; Loghavi, S; Zuniga, LA; Ho, PP; Sobel, RA; Butcher, EC. Chemokine-like receptor-1 expression by central nervous system-infiltrating leukocytes and involvement in a model of autoimmune demyelinating disease. *J Immunol,* 2009, 183, 6717-6723.

[58] Trebst, C; Staugaitis, SM; Kivisakk, P; Mahad, D; Cathcart, MK; Tucky, B; Wei, T; Rani, MR; Horuk, R; Aldape, KD; Pardo, CA; Lucchinetti, CF; Lassmann, H; Ransohoff, RM. CC chemokine receptor 8 in the central nervous system is associated with phagocytic macrophages. *Am J Pathol,* 2003, 162, 427-438.

[59] Vroon, A; Heijnen, CJ; Kavelaars, A. GRKs and arrestins, regulators of migration and inflammation. *J Leukoc Biol,* 2006, 80, 1214-1221.

[60] Thelen, M; Stein, JV. How chemokines invite leukocytes to dance. *Nat Immunol,* 2008, 9, 953-959.

[61] Sallusto, F; Baggiolini, M. Chemokines and leukocyte traffic. *Nat Immunol,* 2008. 9, 949-952.

[62] Viola, A; Molon, B; Contento, RL. Chemokines: coded messages for T-cell missions. *Front Biosci,* 2008, 13, 6341-6353.

[63] Aragay, AM; Ruiz-Gomez, A; Penela, P; Sarnago, S; Elorza, A; Jimenez-Sainz, MC; Mayor, F. Jr; G protein-coupled receptor kinase 2 (GRK2): mechanisms of regulation and physiological functions. *FEBS Lett,* 1998, 430, 37-40.

[64] Pitcher, JA; Freedman, NJ; Lefkowitz, RJ. G protein-coupled receptor kinases. *Annu Rev Biochem,* 1998, 67, 653-692.

[65] Ribas, C; Penela, P; Murga, C; Salcedo, A; Garcia-Hoz, C; Jurado-Pueyo, M; Aymerich, I; Mayor, F. Jr; The G protein-coupled receptor kinase (GRK) interactome, role of GRKs in GPCR regulation and signaling. *Biochim Biophys Acta,* 2007, 1768, 913-922.

[66] De Blasi, A; Parruti, G; Sallese, M. Regulation of G protein-coupled receptor kinase subtypes in activated T lymphocytes. Selective increase

of beta-adrenergic receptor kinase 1 and 2. *J Clin Invest,* 1995, 95, 203-210.

[67] Ramer-Quinn, DS; Baker, RA; Sanders, VM; Activated, T. helper, 1 and T helper 2 cells differentially express the beta-2-adrenergic receptor, a mechanism for selective modulation of T helper 1 cell cytokine production. *J Immunol,* 1997, 159, 4857-4867.

[68] Vroon, A; Kavelaars, A; Limmroth, V; Lombardi, MS; Goebel, MU; Van Dam, AM; Caron, MG; Schedlowski, M; Heijnen, CJ; G protein-coupled receptor kinase 2 in multiple sclerosis and experimental autoimmune encephalomyelitis. *J Immunol,* 2005, 174, 4400-4406.

[69] Vroon, A; Heijnen, CJ; Lombardi, MS; Cobelens, PM; Mayor, F; Jr; Caron, MG; Kavelaars, A. Reduced GRK2 level in T cells potentiates chemotaxis and signaling in response to CCL4. *J Leukoc Biol,* 2004, 75, 901-909.

[70] Oppermann, M; Mack, M; Proudfoot, AE; Olbrich, H. Differential effects of CC chemokines on CC chemokine receptor 5 (CCR5) phosphorylation and identification of phosphorylation sites on the CCR5 carboxyl terminus. *J Biol Chem,* 1999, 274, 8875-8885.

[71] Balabanian, K; Levoye, A; Klemm, L; Lagane, B; Hermine, O; Harriague, J; Baleux, F; Arenzana-Seisdedos, F; Bachelerie, F. Leukocyte analysis from WHIM syndrome patients reveals a pivotal role for GRK3 in CXCR4 signaling. *J Clin Invest,* 2008, 118, 1074-1084.

[72] Lombardi, MS; Kavelaars, A; Schedlowski, M; Bijlsma, JW; Okihara, KL; Van de Pol, M; Ochsmann, S; Pawlak, C; Schmidt, RE; Heijnen, CJ. Decreased expression and activity of G-protein-coupled receptor kinases in peripheral blood mononuclear cells of patients with rheumatoid arthritis. *Faseb J,* 1999, 13, 715-725.

[73] Vroon, A; Lombardi, MS; Kavelaars, A; Heijnen, CJ. Changes in the G-protein-coupled receptor desensitization machinery during relapsing-progressive experimental allergic encephalomyelitis. *J Neuroimmunol,* 2003, 137, 79-86.

[74] Giorelli, M; Livrea, P; Defazio, G; Iacovelli, L; Capobianco, L; Picascia, A; Sallese, M; Martino, D; Aniello, MS; Trojano, M; De Blasi, A. Interferon beta-1a counteracts effects of activation on the expression of G-protein-coupled receptor kinases 2 and 3, beta-arrestin-1, and regulators of G-protein signalling 2 and 16 in human mononuclear leukocytes. *Cell Signal,* 2002, 14, 673-678.

[75] Lombardi, MS; Kavelaars, A; Penela, P; Scholtens, EJ; Roccio, M; Schmidt, RE; Schedlowski, M; Mayor, F; Jr; Heijnen, CJ. Oxidative

stress decreases G protein-coupled receptor kinase 2 in lymphocytes via a calpain-dependent mechanism. *Mol Pharmacol,* 2002, 62, 379-388.

[76] Seya, T; Hirano, A; Matsumoto, M; Nomura, M; Ueda, S; Human membrane cofactor protein (MCP, CD46), multiple isoforms and functions. *Int J Biochem Cell Biol,* 1999, 31, 1255-1260.

[77] Cattaneo, R. Four viruses, two bacteria, and one receptor: membrane cofactor protein (CD46) as pathogens' magnet. *J Virol,* 2004, 78, 4385-4388.

[78] Riley-Vargas, RC; Gill, DB; Kemper, C; Liszewski, MK; Atkinson, JP. CD46, expanding beyond complement regulation. *Trends Immunol,* 2004, 25, 496-503.

[79] Russell, SM; Loveland, BE; Johnstone, RW; Thorley, BR; McKenzie, IF. Functional characterisation of alternatively spliced CD46 cytoplasmic tails. *Transplant Proc,* 1992, 24, 2329-2330.

[80] Johnstone, RW; Russell, SM; Loveland, BE; McKenzie, IF. Polymorphic expression of CD46 protein isoforms due to tissue-specific RNA splicing. *Mol Immunol,* 1993, 30, 1231-1241.

[81] Rudd, CE; Schneider, H. Unifying concepts in CD28, ICOS and CTLA4 co-receptor signalling. *Nat Rev Immunol,* 2003, 3, 544-556.

[82] Sharpe, AH; Freeman, GJ. The B7-CD28 superfamily. *Nat Rev Immunol,* 2002, 2, 116-126.

[83] Jimenez-Perianez, A; Ojeda, G; Criado, G; Sanchez, A; Pini, E; Madrenas, J; Rojo, J. M; Portoles, P. Complement regulatory protein Crry/p65-mediated signaling in T lymphocytes, role of its cytoplasmic domain and partitioning into lipid rafts. *J Leukoc Biol* 2005, 78, 1386-1396.

[84] Gaglia, JL; Mattoo, A; Greenfield, EA; Freeman, GJ; Kuchroo, VK; Characterization of endogenous Chinese hamster ovary cell surface molecules that mediate T cell costimulation. *Cell Immunol,* 2001, 213, 83-93.

[85] Fernandez-Centeno, E; de Ojeda, G; Rojo, JM; Portoles, P; Crry/p65, a membrane complement regulatory protein, has costimulatory properties on mouse T cells. *J Immunol,* 2000, 164, 4533-4542.

[86] Morgan, BP; Marchbank, KJ; Longhi, MP; Harris, CL; Gallimore, AM. Complement: central to innate immunity and bridging to adaptive responses. *Immunol Lett,* 2005, 97, 171-179.

[87] Kemper, C; Atkinson, JP. T-cell regulation: with complements from innate immunity. *Nat Rev Immunol,* 2007, 7, 9-18.

[88] Longhi, MP; Harris, CL; Morgan, BP; Gallimore, A; Holding, T. cells in check--a new role for complement regulators? *Trends Immunol,* 2006, 27, 102-108.

[89] Sanchez, A; Feito, MJ; Rojo, JM. CD46-mediated costimulation induces a Th1-biased response and enhances early TCR/CD3 signaling in human CD4+ T lymphocytes. *Eur J Immunol,* 2004, 34, 2439-2448.

[90] Oliaro, J; Pasam, A; Waterhouse, NJ; Browne, KA; Ludford-Menting, MJ; Trapani, JA; Russell, SM. Ligation of the cell surface receptor, CD46, alters T cell polarity and response to antigen presentation. *Proc Natl Acad Sci,* U S A, 2006, 103, 18685-18690.

[91] Russell, S. CD46: a complement regulator and pathogen receptor that mediates links between innate and acquired immune function. *Tissue Antigens,* 2004, 64, 111-118.

[92] Lee, SW; Bonnah, RA; Higashi, DL; Atkinson, JP; Milgram, SL; So, M. CD46 is phosphorylated at tyrosine 354 upon infection of epithelial cells by Neisseria gonorrhoeae. *J Cell Biol,* 2002, 156, 951-957.

[93] Kurita-Taniguchi, M; Hazeki, K; Murabayashi, N; Fukui, A; Tsuji, S; Matsumoto, M; Toyoshima, K; Seya, T. Molecular assembly of CD46 with CD9, alpha3-beta1 integrin and protein tyrosine phosphatase SHP-1 in human macrophages through differentiation by GM-CSF. *Mol Immunol,* 2002, 38, 689-700.

[94] Karp, CL; Wysocka, M; Wahl, LM; Ahearn, JM; Cuomo, PJ; Sherry, B; Trinchieri, G; Griffin, DE. Mechanism of suppression of cell-mediated immunity by measles virus [published erratum appears in Science 1997 Feb 21;275(5303), 1053]. *Science,* 1996, 273, 228-231.

[95] Schnorr, JJ; Xanthakos, S; Keikavoussi, P; Kampgen, E; ter Meulen, V; Schneider-Schaulies, S. Induction of maturation of human blood dendritic cell precursors by measles virus is associated with immunosuppression. *Proc Natl Acad Sci,* U S A, 1997, 94, 5326-5331.

[96] Smith, A; Santoro, F; Di Lullo, G; Dagna, L; Verani, A; Lusso, P. Selective suppression of IL-12 production by human herpesvirus 6. *Blood,* 2003, 102, 2877-2884.

[97] Wong, TC; Yant, S; Harder, BJ; Korte-Sarfaty, J; Hirano, A. The cytoplasmic domains of complement regulatory protein CD46 interact with multiple kinases in macrophages. *J Leukoc Biol,* 1997, 62, 892-900.

[98] Ludford-Menting, MJ; Thomas, SJ; Crimeen, B; Harris, LJ; Loveland, BE; Bills, M; Ellis, S; Russell, SM. A functional interaction between CD46 and DLG4: a role for DLG4 in epithelial polarization. *J Biol Chem,* 2002, 277, 4477-4484.

[99] Hirano, A; Yang, Z; Katayama, Y; Korte-Sarfaty, J; Wong, TC. Human CD46 enhances nitric oxide production in mouse macrophages in response to measles virus infection in the presence of gamma interferon: dependence on the CD46 cytoplasmic domains [In Process Citation]. *J Virol,* 1999, 73, 4776-4785.

[100] Wang, G; Liszewski, M; Chan, A; Atkinson, J. Membrane cofactor protein (MCP; CD46): isoform-specific tyrosine phosphorylation. *J Immunol,* 2000, 164, 1839-1846.

[101] Joubert, PE; Meiffren, G; Gregoire, IP; Pontini, G; Richetta, C; Flacher, M; Azocar, O; Vidalain, PO; Vidal, M; Lotteau, V; Codogno, P; Rabourdin-Combe, C; Faure, M. Autophagy induction by the pathogen receptor CD46. *Cell Host Microbe,* 2009, 6, 354-366.

[102] Shusta, EV; Zhu, C; Boado, RJ; Pardridge, WM. Subtractive expression cloning reveals high expression of CD46 at the blood-brain barrier. *J Neuropathol Exp Neurol,* 2002, 61, 597-604.

[103] Johansson, L; Rytkonen, A; Bergman, P; Albiger, B; Kallstrom, H; Hokfelt, T; Agerberth, B; Cattaneo, R; Jonsson, AB. CD46 in meningococcal disease. *Science,* 2003, 301, 373-375.

[104] Meeuwsen, S; Persoon-Deen, C; Bsibsi, M; Bajramovic, JJ; Ravid, R; De Bolle, L; van Noort, JM. Modulation of the cytokine network in human adult astrocytes by human herpesvirus-6A. *J Neuroimmunol,* 2005, 164, 37-47.

[105] Noe, KH; Cenciarelli, C; Moyer, SA; Rota, PA; Shin, ML. Requirements for measles virus induction of RANTES chemokine in human astrocytoma-derived U373 cells. *J Virol* 1999. 73, 3117-3124.

[106] Soldan, SS; Fogdell-Hahn, A; Brennan, MB; Mittleman, BB; Ballerini, C; Massacesi, L; Seya, T; McFarland, HF; Jacobson, S. Elevated serum and cerebrospinal fluid levels of soluble human herpesvirus type 6 cellular receptor, membrane cofactor protein, in patients with multiple sclerosis. *Ann Neurol,* 2001, 50, 486-493.

[107] Fogdell-Hahn, A; Soldan, SS; Shue, S; Akhyani, N; Refai, H; Ahlqvist, J; Jacobson, S. Co-purification of soluble membrane cofactor protein (CD46) and human herpesvirus 6 variant A genome in serum from multiple sclerosis patients. *Virus Res,* 2005, 110, 57-63.

[108] Tuke, PW; Hawke, S; Griffiths, PD; Clark, DA. Distribution and quantification of human herpesvirus 6 in multiple sclerosis and control brains. *Mult Scler,* 2004, 10, 355-359.

[109] Astier, AL; Hafler, DA. Abnormal Tr1 differentiation in multiple sclerosis. *J Neuroimmunol,* 2007, 191, 70-78.

[110] Ma, A; Xiong, Z; Hu, Y; Qi, S; Song, L; Dun, H; Zhang, L; Lou, D; Yang, P; Zhao, Z; Wang, X; Zhang, D; Daloze, P; Chen, H. Dysfunction of IL-10-producing type 1 regulatory T cells and CD4+CD25+ regulatory T cells in a mimic model of human multiple sclerosis in Cynomolgus monkeys. *International Immunopharmacology,* 2009, 9, 599-608.

[111] Kivisakk, P; Tucky, B; Wei, T; Campbell, JJ; Ransohoff, RM. Human cerebrospinal fluid contains CD4+ memory T cells expressing gut- or skin-specific trafficking determinants: relevance for immunotherapy. *BMC Immunol,* 2006, 7, 14.

[112] Yednock, TA; Cannon, C; Fritz, LC; Sanchez-Madrid, F; Steinman, L; Karin, N. Prevention of experimental autoimmune encephalomyelitis by antibodies against alpha 4 beta 1 integrin. *Nature,* 1992, 356, 63-66.

[113] Brocke, S; Piercy, C; Steinman, L; Weissman, IL; Veromaa, T. Antibodies to CD44 and integrin alpha4, but not L-selectin, prevent central nervous system inflammation and experimental encephalomyelitis by blocking secondary leukocyte recruitment. *Proc Natl Acad Sci,* U S A 1999, 96, 6896-6901.

[114] Kanwar, JR; Harrison, JE; Wang, D; Leung, E; Mueller, W; Wagner, N; Krissansen, GW. Beta7 integrins contribute to demyelinating disease of the central nervous system. *J Neuroimmunol,* 2000, 103, 146-152.

[115] Davenport, RJ; Munday, JR. Alpha4-integrin antagonism-an effective approach for the treatment of inflammatory diseases? *Drug Discov Today,* 2007, 12, 569-576.

[116] Mellergard, J; Edstrom, M; Vrethem, M; Ernerudh, J; Dahle, C. Natalizumab treatment in multiple sclerosis: marked decline of chemokines and cytokines in cerebrospinal fluid. *Mult Scler.*, 16, 208-217.

In: Chemotaxis: Types, Clinical Significance... ISBN: 978-1-61728-495-3
Editor: T. C. Williams, pp. 109-133

Chapter 4

Regulation of Chemotaxis by Heterotrimeric G Proteins

***Maggie M. K. Lee*[1] *and Yung H. Wong*[2*]**
Department of Biochemistry, the Molecular Neuroscience Center, and the Biotechnology Research Institute, Hong Kong University of Science and Technology, Clear Water Bay, Kowloon, Hong Kong, China

Abstract

With the ability to mediate chemotaxis of inflammatory cells, chemoattractants have been shown to contribute to the development of inflammatory diseases, such as atherosclerosis and angiogenesis. Many chemoattractant receptors belong to the family of seven-transmembrane-domain G protein-coupled receptors (GPCRs) and elicit their effects through heterotrimeric (αβγ) guanine nucleotide-binding proteins (G proteins). The three subunits form two functional compartments – the Gα subunit and the stable Gβγ complex. Both the dissociated GTP-bound Gα subunit and Gβγ complex can exert biological effects. G proteins are classified into four major subfamilies, G_s, G_i, G_q and G_{12}, according to the amino acid sequence homology and functional specialization of the Gα subunit. Members in all G protein subfamilies are known to interact with

* Corresponding author: Department of Biochemistry, Hong Kong University of Science and Technology, Clear Water Bay, Kowloon, Hong Kong, China. Tel: (852) 2358 7328 Fax: (852) 2358 1552, Email: boyung@ust.hk

chemoattractant receptors individually or simultaneously, which trigger the activation of multiple signaling molecules, leading to actin reorganization and subsequent cell mobilization. In this chapter, we will discuss the promiscuity of chemoattractant receptors in G protein coupling and examine their underlying molecular mechanisms in directed cell migration.

Introduction

Chemotaxis is a phenomenon in which the direction of cell migration is determined by an extracellular gradient of chemicals. This directional cell mobilization plays a critical role in many diverse physiological processes, including the recruitment of leukocytes to sites of inflammation, trafficking of lymphocytes from bone marrow to secondary lymphoid organs, and migration of vascular smooth muscle cells for angiogenesis in menstrual cycle. Dysregulation of chemotaxis can cause diverse pathological conditions, including tumor growth [Koizumi *et al.*, 2007] and inflammatory diseases such as asthma, arthritis and atherosclerosis [Liehn *et al.*, 2006; Silva *et al.*, 2007].

Many molecular components involved in chemotaxis of eukaryotic cells have been discovered. In mammals, the extracellular stimuli that guide cell migration include cell-secreted proteins (chemokine), bacterial peptides (N-formyl methionyl leucyl phenylalanine/fMLP), and products of phospholipid metabolism. Platelet-activating factor (PAF), sphingosine 1-phosphate (S1P), lysophosphatidic acid (LPA), lysophosphatidylcholine (LPC) and leukotrienes (LTB_4) are bioactive lipids capable of mediating cell mobilization. Chemokines are composed of 70 to 125 amino acids [Olson and Ley, 2002]. Most of them contain four conserved cysteines that form two disulfide bonds, one between the first and the third cysteines while another between the second and the fourth cysteines. Chemokines are classified into four subfamilies according to the number of amino acids between the first two conserved cysteines. They include CXC chemokines such as CXCL8/IL-8 and CXCL12/SDF-1, CC chemokines such as CCL2/MCP-1, C chemokines named as XCL1/lymphotactin and XCL2/SCM-1β, and CX_3C chemokine named as CX_3CL1/fractalkine [Ono *et al.*, 2003]. All chemokines exert their effects via chemokine receptors, which are G protein-coupled receptors (GPCRs) with seven transmembrane domains. Up till now, nineteen chemokine receptors have been identified. They are classified into four subfamilies depending on their chemokine specificity [Olson and Ley, 2002]. S1P is a lysophospholipid mediator produced exclusively by sphingosine kinase (SPHK) 1 and SPHK2 *in vivo*. S1P receptors contain five subtypes and S1P interacts mainly with $S1P_1$, $S1P_2$ and $S1P_3$ [Radeff-Huang *et al.*, 2004]. LPA is generated by the conversion of LPC by LPLD/autotoxin [Mills and Moolenaar, 2003] or the hydrolysis of phosphatidic acid by phospholipase A_2 (PLA_2) [Maghazachi, 2005]. In mammals, four LPA receptors ($LPA_{1\text{-}4}$) have been identified [Radeff-Huang *et al.*, 2004; Ye *et al.*, 2002]. S1P and LPA, as well as PAF and LTB_4, signal through their respective GPCR subtypes and the cellular response depends upon the cell type and/or the cellular context (Table 1).

GPCRs exert their effects through heterotrimeric guanine nucleotide-binding proteins (G proteins). Heterotrimeric G proteins are made up of three subunits: Gα, Gβ and Gγ. The three subunits form two functional compartments – the Gα subunit and the stable Gβγ complex. Gα subunits belong to a group of enzymes called GTP hydrolases [GTPase; Bourne *et al.*, 1991]. At the resting stage, the GDP-bound Gα subunit is associated with Gβγ complex. Upon ligand binding, GPCR acts as a guanine nucleotide exchange factor (GEF) for the Gα subunit, causing its release of GDP and binding of GTP. The conformational changes of GTP-bound Gα subunit reduce its affinity for the Gβγ complex. Both the dissociated GTP-bound Gα subunit and Gβγ complex can exert biological effects. The intrinsic GTPase activity of Gα subunit leads to GTP hydrolysis and restores it back to the GDP-bound form to associate with the Gβγ complex again. The Gα subunit is composed of several important functional domains in charge of GTP binding and hydrolysis, receptor recognition, effector regulation, Gβγ subunit binding, membrane attachment and modification by bacterial toxins [Simon *et al.*, 1991; Spiegel, 1992].

G proteins are classified into four major subfamilies according to the amino acid sequence homology and functional specialization of the Gα subunit [Simon *et al.*, 1991]. The four subfamilies are: G_s, which activates adenylyl cyclase (AC) and Ca^{2+} channel but inhibits Na^+ channel; G_i, which inhibits AC and Ca^{2+} channel but stimulates K^+ channel; G_q, which activates phospholipase Cβ (PLCβ); and G_{12}, which regulates cell growth and differentiation as well as Na^+/H^+ ion exchange. Members in all G protein subfamilies are known to interact with chemoattractant receptors individually or simultaneously, which trigger the action of multiple signaling molecules, leading to actin reorganization and subsequent cell migration. This chapter attempts to review the regulation of chemotaxis by the four subfamilies of Gα subunits and their associated Gβγ complex.

Table 1. Signaling requirements for chemotaxis

Ligand/ Receptor	Cell type	G_i	G_q	G_{12}	Reference
G_i, G_q or G_{12}					
CCL19	mouse T cells	✓	X	NA	Shi *et al.*, 2007
CXCL8	mouse BM neutrophils	✓	X	NA	Shi *et al.*, 2007
CXCL12	mouse T cells	✓	X	NA	Shi *et al.*, 2007
PAF	rat basophilic leukemic RBL-2H3 cells	✓	X	NA	Brown *et al.*, 2006
fMLP	rat RBL-2H3 cells	✓	X	NA	Haribabu *et al.*, 1999
LTB_4	rat RBL-2H3 cells	✓	X	NA	Haribabu *et al.*, 1999
$S1P/S1P_1$	transfected CHO cells	✓	NA	NA	Sugimoto *et al.*, 2003
$S1P/S1P_2$	transfected CHO cells	X	X	✓	Sugimoto *et al.*, 2003
oxytocin	human HUVECs	X	✓	NA	Cattaneo *et al.*, 2008
G_i and G_q					
CCL2	human monocytic THP-1 cells	✓	✓ ($G\alpha_{16}$)	NA	Tian *et al.*, 2008
CCL3	mouse BM neutrophils	✓	✓	NA	Shi *et al.*, 2007
CCL15	human monocytic THP-1 cells	✓	✓ ($G\alpha_{16}$)	NA	Tian *et al.*, 2008
CCL19	mouse DC	✓	✓	NA	Shi *et al.*, 2007
CXCL10	human IL-2 activated NK cells	✓	✓	NA	Maghazachi *et al.*, 1994
CXCL12	mouse DC	✓	✓	NA	Shi *et al.*, 2007
XCL1	human IL-2 activated NK cells	✓	✓	NA	Maghazachi *et al.*, 1994
fMLP	mouse BM neutrophils	✓	✓	NA	Shi *et al.*, 2007
LPA	mouse and rat VSMCs	✓	✓	NA	Kim *et al.*, 2006
G_i and G_{12}					
CXCL12	human Jurkat T cells	✓	NA	✓ ($G\alpha_{13}$)	Tan *et al.*, 2006
$S1P/S1P_3$	transfected CHO cells	✓	NA	✓	Sugimoto *et al.*, 2003
LPA	human ovarian cancer SK-OV3 cells	✓	X	✓	Bian *et al*, 2006
G_q and G_{12}					
$S1P/S1P_2$	rat VSMCs	X	✓	✓	Takashima *et al.*, 2008
LPC	mouse macrophages J774A.1	X	✓	✓	Yang *et al.*, 2005

BM: Bone marrow, DC: dendritic cells; HUVECs: human umbilical vein endothelial cells; NA: not available; NK: natural killer; VSMCs: vascular smooth muscle cells

G_i Subfamily

G_i subfamily is the largest subfamily of G proteins generally known as "inhibitory" $G\alpha$ subunits. G_i subfamily has nine members. They are G_{i1-3}, G_{oA-B}, G_{t1-2}, G_z and G_{gust}. All $G\alpha$ subunits of G_i proteins, except those of G_z, contain the cysteine residue four amino acids from the C-terminus that is the target for NAD^+-dependent ADP-ribosylation. The ADP-ribosylation is catalyzed by pertussis toxin (PTX) and this modification prevents the activation of G_i proteins by receptor. PTX-sensitivity becomes a signature of the involvement of G_i class proteins in signaling systems [Simon *et al.*, 1991]. $G\alpha_i$ subunits are known to inhibit AC. ACs are a family of membrane bound enzymes that catalyze the formation of cAMP from ATP. Many chemoattractant receptors are G_i-coupled receptors. It has been shown that in HEK293 cells stably transfected with CCR2b, CCL2 induces AC inhibition which is sensitive to PTX pretreatment [Myers *et al.*, 1995]. In NG108-15 cells expressing CCR5, CCL5/RANTES stimulates [^{35}S]GTPγS binding to cell membranes and induces inhibition of AC activity. The AC inhibition by CCL5 is PTX-sensitive and overexpression of $G\alpha_{i2}$ strongly increases the inhibition on AC [Zhao *et al.*, 1998]. It has also been reported by different groups that CXCL8 can mediate AC inhibition in CHO cells stably transfected with CXCR1 or CXCR2. The CXCL8-induced AC inhibition is PTX-sensitive in both cell types [Hall *et al.*, 1999; Shyamala and Khoja, 1998]. In platelets endogenously expressing CXCR4, CXCL12 inhibits cAMP formation [Kowalska *et al.*, 2000]. These findings suggest that CCR2b, CCR5, CXCR1, CXCR2 and CXCR4 are functionally coupled to G_i proteins.

PAF receptor couples to both PTX-sensitive and PTX-insensitive G proteins to mediate cellular responses. Reconstitution studies in COS-7 cells show that PAF receptor can couple to $G\alpha_q$, $G\alpha_{11}$ and $G\alpha_{16}$ to activate PLCβ [Amatruda *et al.*, 1993]. However, PAF receptor uses $G\alpha_{i3}$, but not $G\alpha_q$, to mediate the migration of rat basophilic leukemia cells (RBL-2H3) [Brown *et al.*, 2006]. Similarly, fMLP and LTB_4-stimulated chemotaxis of rat RBL-2H3 cells is regulated by G_i proteins [Haribabu *et al.*, 1999]. CC chemokine receptors, CCR1, CCR2b, CCR3, are capable of mediating chemokine-induced PLCβ stimulation via either G_{14} or G_{16} [Tian *et al.*, 2008]. Likwise, CXC chemokine receptors, CXCR1 and CXCR2, are coupled to G_{16} [Kuang *et al.*, 1996; Wu *et al.*, 1993]. Nevertheless, CCR1 agonist (CCL15/Lkn-1)-induced chemotaxis has been demonstrated to be PTX-sensitive in human osteogenic sarcoma (HOS) cells overexpressing CCR1, indicating that $G_{i/o}$ proteins are involved in cell migration [Ko *et al.*, 2002]. The CCL15-mediated migratory response requires PLC and protein kinase C (PKC)γ. Other examples of chemokine receptors utilizing G_i protein in cell mobilization include CCR2 [Cambien *et al.*, 2001; Sozzani *et al.*, 1991], CCR3 [Ponath *et al.*, 1996], CXCR1, CXCR2 [Sebok *et al.*, 1993] and CXCR4 [Kim *et al.*, 2001; Soriano *et al.*, 2003; Vila-Coro *et al.*, 1999]. $G\alpha_{i2}$ and $G\alpha_{i3}$ proteins may play distinct, antagonizing and additive roles depending on the specific receptor. For example, $G\alpha_{i2}$ is indispensable for the migratory responses of T cells to three CXCR3 ligands, CXCL9/Mig, CXCL10/IP-10 and CXCL11/I-TAC, as the lack of $G\alpha_{i2}$ abolishes CXCR3-stimulated migration. In contrast, T cells isolated from $G\alpha_{i3}$ knockout mice displayed a significant increase in chemotaxis when stimulated with CXCR3 agonists [Thompson *et al.*, 2007]. On the contrary, the CXCR4 receptor requires both $G\alpha_{i2}$ and $G\alpha_{i3}$ for a full response of cell mobilization [Thompson *et al.*, 2007].

$S1P_1$-deficient T cells fail to exit from thymus and therefore are completely absent from the peripheral bloodstream [Matloubian *et al.*, 2004]. The defective vascular maturation observed in $S1P_1$-deficient mice highlights a fundamental role for S1P signaling on vasculogenesis [Liu *et al.*, 2000]. Apart from having potent migratory effects on $S1P_1$ expressing vascular endothelial cells [Lee *et al.*, 1999; Wang *et al.*, 1999], S1P has been demonstrated to induce the migration and capillary-like tube formation of human lymphatic endothelial cells (HLECs) that endogenously expressed $S1P_1$ [Yoon *et al.*, 2008]. The $S1P_1$-mediated response is completely blocked by PTX and is partially inhibited by U73122 and BAPTA-AM, showing the involvement of G_i protein, PLC and Ca^{2+}. $S1P_1$ is coupled exclusively to the G_i protein family, whereas $S1P_2$ and $S1P_3$ are coupled to the G_i, G_q and $G_{12/13}$ protein families [Taha *et al.*, 2004]. In contrast to $S1P_2$, $S1P_1$ and $S1P_3$ mediated S1P-directed chemotaxis and Rac activation via $G\alpha_i$ [Sugimoto *et al.*, 2003]. S1P mediates either stimulatory or inhibitory regulation for cell migration. This bimodal regulation of chemotaxis by S1P is based upon a diversity of S1P receptor isotypes. $S1P_2$ acts as a repellant receptor to mediate inhibition of chemotaxis towards attractants, whereas $S1P_1$ and $S1P_3$ act as attractant receptors to mediate migration directed towards S1P [Okamoto *et al.*, 2000].

LPA receptors have been linked to three major G proteins, G_i, G_q and $G_{12/13}$ [Siess, 2002]. Calcium mobilization induced by LPA in LPA_1-transfected cells is maintained through PTX-sensitive $G\alpha_i$ only, whereas the same response is mediated by PTX-sensitive $G\alpha_i$ and PTX-insensitive $G\alpha_q$

upon activation of LPA_2 [An *et al.*, 1998]. LPA activates Rho and induces cytoskeleton rearrangement through $G\alpha_{12/13}$ [Kranenburg *et al.*, 1999]. LPA/LPA_2 stimulates migration of human ovarian cancer CAOV-3 cells via extracellular signal-regulated kinase (ERK) and cyclooxygenase-2 (COX-2) [Jeong *et al.*, 2008]. The LPA-induced COX-2 expression is inhibited by PTX as well as by inhibitors of Src, epidermal growth factor receptor (EGFR) and ERK [Jeong *et al.*, 2008]. Although chemoattractant receptors are primarily coupled to G_i proteins for chemotaxis, it is believed that the release of $G\beta\gamma$ complex, rather than $G\alpha_i$ subunits themselves, contributes to chemotaxis [Arai *et al.*, 1997; Neptune and Bourne, 1997]. The effect of $G\beta\gamma$ on chemotaxis will be discussed in a later section.

G_s Subfamily

G_s proteins are classified into $G\alpha_{sL}$, $G\alpha_{sS}$ and $G\alpha_{solf}$. All $G\alpha_s$ proteins are sensitive to cholera toxin (CTX). CTX catalyzes the NAD^+-dependent ADP-ribosylation of an arginine residue in these $G\alpha$ subunits. This modification inhibits the GTPase activity of G_s proteins and locks their α subunits in the GTP-bound active conformations. Within this subfamily, $G\alpha_{sL}$ and $G\alpha_{sS}$ are ubiquitous and they stimulate AC and Ca^{2+} channels but inhibit Na^+ channels. In contrast, $G\alpha_{solf}$ is mainly expressed in olfactory neuroepithelium and activates AC. Activation of AC results in increases in cAMP, leading to activation of cAMP-dependent protein kinase (PKA) and subsequent phosphorylation of cellular proteins [Sadana and Dessauer, 2009].

Prostaglandin E2 (PGE_2), via stimulation of the EP2 receptor, promotes human squamous cell carcinoma growth and invasion through a mechanism involving activation of PKA, ERK and inducible NO synthase (iNOS)/guanylate cyclase (GC) pathway [Donnini *et al.*, 2007]. The PGE_2-induced iNOS activation depends on Src, PKA and EGFR. Activation of EP4 receptors by PGE_2 also participates in regulating cell migration. Using i.v. injected Lewis lung carcinoma (3LL), it has been found that tumor metastasis to lung is significantly reduced when mice are treated with a specific EP4 antagonist or when EP4 receptor expression is knocked down in the tumor cells using RNA interference technology [Yang *et al.*, 2006]. There is only sporadic evidence to suggest that G_s proteins can regulate the chemotactic response of leukocytes. For example, CCL8/MCP2-induced chemotaxis of human monocytes [Sozzani *et al.*, 1994] and CCL3/MIP-1α-stimulated chemotaxis of human IL-2-activated natural killer (NK) cells [Maghazachi *et al.*, 1994] are unaffected by PTX whereas they are inhibited by CTX pretreatment, suggesting the involvement of G_s proteins. However, the signaling molecules involved in G_s-coupled receptor-mediated chemotaxis of leukocytes remain to be determined. Nevertheless, $G\alpha_s$ has been demonstrated to be required in homing of hematopoietic stem and progenitor cells (HSPCs) to the bone marrow [Adams *et al*, 2009].

G_q Subfamily

G_q subfamily contains five members, including G_q, G_{11}, G_{14} and $G_{15/16}$, with G_{15} being the mouse homolog of human G_{16}. These proteins are resistant to ADP-ribosylation by either PTX or CTX. Unlike the ubiquitous expression of G_q and G_{11}, the distributions of G_{14} and G_{16} are more tissue-specific. G_{14} is predominantly expressed in pancreatic islets, taste tissue, spleen, lung, kidney, testis and hematopoietic cells whereas G_{16} is restricted to hematopoietic cells [Amatruda *et al.*, 1991; McLaughlin *et al.*, 1994; Nakamura *et al.*, 1991; Wilkie *et al*, 1991; Zigman *et al.*, 1994]. Both G_{14} and G_{16} are more promiscuous than other G proteins in terms of receptor recognition; they can bind to a variety of GPCRs including some receptors which are typically characterized as G_i-coupled receptors [Ho *et al.*, 2001; Lee *et al.*, 1998, Offermanns and Simon, 1995; Su *et al.*, 2009].

GPCRs utilize G_q proteins for the activation of PLCβ. PLC isozymes are classified into three families: PLCβ, PLCγ and PLCδ. All Gα subunits of G_q class proteins can activate β isoforms of PLC [Jiang *et al.*, 1994; Lee *et al.*, 1992; Mizuno and Itoh, 2009]. PLCβ catalyzes the hydrolysis of phosphatidylinositol 4,5-bisphosphate (PIP_2) into two important intracellular second messengers, inositol 1,4,5-trisphosphate (IP_3) and diacylglycerol (DAG) [Berridge and Irvine, 1989]. The IP_3 will then induce calcium release from endoplasmic reticulum while the DAG will activate PKC to trigger a cascade of phosphorylation reactions [Walsh *et al.*, 1994].

Chemokine receptors primarily utilize the PTX-sensitive G_i proteins for chemotaxis. However, chemokines can induce cell movement through a PTX-insensitive pathway. It has been shown that CCL2-induced chemotaxis are only partially inhibited by PTX in a variety of cell types, including human monocytes [Yen *et al.*, 1997], T cells [Sotsios et al., 1999], B cells [Frade *et al.*, 1997] and IL2-activated NK cells [Allavena *et al.*, 1994; Maghazachi *et al.*, 1994]. The involvement of Gα_{16} has been recently demonstrated in CCL15-stimulated migration of human monocytic THP-1 cells [Tian *et al.*, 2008]. Despite the inactivation of $G_{i/o}$ proteins by PTX, CCL15 remains effective as a chemotactic agent. Similar resistance to PTX treatment has been observed for CCL2. In addition, CCL15 and CCL2 are less efficacious in stimulating chemotaxis of THP-1 cells following the knockdown of Gα_{16} by overexpressing siRNA, indicating the participation of Gα_{16} in CCR1 or CCR2b-mediated cell migration [Tian *et al.*, 2008].

Oxytosin (OT) may either inhibit or stimulate cell migration. It has been found to inhibit the migration of ovarian cancer cells [Morita *et al.*, 2004], but it promotes the migration of human umbilical vein endothelial cells (HUVECs) [Cattaneo *et al.* 2008], immortalized human dermal microvascular endothelial cells and breast carcinoma-derived endothelial cells [Cassoni *et al.*, 2006]. OT-mediated migration and invasion of human HUVECs is insensitive to PTX treatment. G_q coupling, activation of PLC and phosphatidylinostiol 3-kinase (PI3K), and formation of nitric oxide (NO) are apparently required for the pro-migratory effect of OT [Cattaneo *et al.* 2008]. The positive or negative regulation of cell migration can also be observed in ATP and UTP-driven chemotaxis. Metabotropic P2Y receptor (P2Y$_2$R) has been reported to promote cell motility of rat aortic smooth muscle cells (SMCs) [Chaulet *et al.*, 2001; Pillois *et al.*, 2002], HUVECs [Taboubi *et al.*, 2007], rabbit corneal epithelial cells [Pintor *et al.*, 2004], rat primary astrocytes [Wang *et al.*, 2005] and human neutrophils [Chen *et al.*, 2006]. In contrast, extracellular ATP and UTP induce via P2Y$_2$R and G$\alpha_{q/11}$, a potent inhibition of human keratinocyte cell spreading and lamellipodium dynamics, and disorganize the actin cytoskeleton and focal contacts [Taboubi *et al.*, 2007].

G_{12} Subfamily

G_{12} subfamily is the smallest subfamily of Gα subunits because it consists of only two members, Gα_{12} and Gα_{13}. Both of them are resistant to CTX or PTX treatment. Members of G_{12} subfamily are responsible for the regulation of c-Jun N-terminal kinase (JNK), Na^+/H^+ ion exchange and focal adhesion assemblies [Dhanasekaran and Dermott, 1996; Hooley *et al.*, 1996; Lin *et al.*, 1996]. Gα_{12} and Gα_{13} have been shown to active Rho GTPase through direct interaction of G$\alpha_{12/13}$ with guanine nucleotide exchange factor (RhoGEFs) [Fukuhara *et al.*, 1999; Fukuhara *et al.*, 2000; Hart *et al.*, 1998; Kozasa *et al.*,

1998; Suzuki *et al.*, 2009]. Rho GTPase, belonging to the Ras superfamily of small monomeric G proteins, is widely known for regulating the actin cytoskeleton and for activating transcription [Hall, 1998; Ridley, 1997].

$G\alpha_{13}$ knockout embryonic fibroblasts exhibit reduced migratory response to thrombin and LPA [Offermanns *et al.*, 1997]. It has been shown that $G\alpha_{13}$ is involved in CXCL12-induced Rho activation and migration of human Jurkat T cells [Tan *et al.*, 2006]. By expression of specific Gα C-terminal peptides, it has been demonstrated that $G\alpha_{12}$ and $G\alpha_{13}$, but not $G\alpha_q$, couple $S1P_2$ to inhibition of Rac, membrane ruffling, and cell migration [Sugimoto *et al.*, 2003].

Gβγ Complex

Gβγ subunits can regulate a number of effectors, including PLCβ2, PLCβ3, PI3Kγ, and isoforms of AC [Neer, 1995]. It has been demonstrated that Gβγ directly interacts with p21-activated kinase 1 (PAK1) and activates Cdc42 through PAK1-associated guanine nucleotide exchange factor (PIXα) [Li *et al.*, 2003]. The recent discovery of P-Rex1, a Gβγ and phosphatidylinositol(3,4,5)triphosphate (PIP_3)-dependent GEF for Rac, suggests that chemoattractants regulate Rac via P-Rex1 [Welch *et al.*, 2002]. Hwang *et al.* have shown that deletion of $G\beta_2$ ablates complement C5a and C3a-provoked migration of mouse macrophages [Hwang *et al.*, 2004].

Activation of chemoattractant receptors by CXCL8, fMLP and C5a stimulates PI3Kγ through the actions of Gβγ and Ras, resulting in the production of PIP_3 in the plasma membrane [Andrews *et al.*, 2007]. The binding of chemoattractant to GPCR induces the dissociation of heterotrimeric G proteins into Gα and Gβγ complex. Free Gβγ complex activates the small GTPase Ras predominantly at leading edge of the cell. Activated Ras subsequently binds to PI3Kγ and begins the rapid conversion of PIP_2 to PIP_3 in the plasma membrane [Bourne and Weiner, 2002]. PIP_3 then recruits PH-domain containing proteins such as Akt/PKB as well as small GTPase Rac and Cdc42 to the cell's leading edge [Stradal *et al.*, 2004; Weiner, 2002]. Activated Rac and Cdc42 interact with the Wiskott-Aldrich syndrome protein (WASP)/Suppressor of cAMP receptor (SCAR) complex leading to actin polarization [Stradal *et al.*, 2004]. Activation of chemoattractant receptors also regulates the localization and activity of 3' phosphatase and tensin homolog (PTEN) by small GTPase RhoA and Cdc42 to modulate PIP_3 levels around the cell membrane [Li *et al.*, 2003]. In addition to the signaling network that predominantly acts at the front of migrating neutrophils, another pathway appears to be operational at the tail end. It has been shown that the fMLP receptor activates G_{12} and G_{13}, instead of G_i, at the back. The activated G_{12} and G_{13} recruit RhoA to the membrane to activate Rho-dependent kinase (ROCK) and myosin II, resulting in the assembly of contractile actomyosin complexes [Wong *et al.*, 2007; Xu *et al.*, 2003].

Signal Integration of G_i and G_q

A further complicating factor in G protein-mediated chemotaxis is the simultaneous activation of multiple G proteins by GPCRs, since the individual signaling pathways do not operate in isolation and they often converge at various loci. Figure 1 illustrates the proposed model of signal integration by G_i and G_q (A), G_i and G_{12} (B), or G_q and G_{12} (C) that eventually lead to cell migration. Moreover, simultaneous activation of GPCRs of the same or different class may also occur. Examples of signal integration are abundant and are not limited to G protein signals alone. CD38 is a nicotinamide adenine dinucleotide (NAD) glycohydrolase and ADP-ribosyl cyclase. CD38-dependent chemoattractants requires both $G\alpha_i$ and $G\alpha_q$ to induce chemotaxis. Bone marrow (BM) neutrophils from $G\alpha_q$-deficient mouse elicit defective chemotactic responses upon fMLP and CCL3 stimulation, whereas CCL19/MIP-3β and CXCL12/SDF-1α require $G\alpha_q$ to mobilize mouse dendritic cells (DC). In contrast, $G\alpha_q$-deficient T cell responses to CXCL12 and CCL19 remain intact [Shi *et al.*, 2007]. The monitoring of fMLP-induced calcium mobilization in mouse bone marrow neutrophils suggests that $G\alpha_{i2}$ controls the IP_3-gated calcium release and that $G\alpha_q$ and CD38 coordinately sustain the calcium response by activating calcium entry [Shi *et al.*, 2007]. Both CXCL10 and XCL1 induce chemotaxis and intracellular calcium mobilization of human IL-2-activated NK cells. Introduction of antibodies to $G\alpha_i$, $G\alpha_o$ or $G\alpha_q$ inhibited both CXCL10 and XCL1-induced migratory responses, showing that G_i and G_q proteins work in concert in chemoattractant-stimulated cell mobilization [Maghazachi *et al.*, 1994]. Likewise, LPA-induced migration of mouse and rat vascular smooth muscle cells (VSMCs) is coupled to both G_i and G_q-mediated mechanisms [Kim *et al.*, 2006]. LPA-induced actin reorganization, which is a fundamental process in cell motility and division, is mediated by both G_i and G_q pathways [Hirshman and Emala, 1999].

Signal Integration of G_i and G_{12}

CXCL12/CXCR4 signaling system is now known to be critical for the regulation of the migration, proliferation, differentiation, and survival of lymphocytes, as reflected by its key role in lymphocyte trafficking and overall immune surveillance [Juarez and Bendall, 2004]. CXCR4 is also an obligatory co-receptor for the infection of T-cell tropic human immunodeficiency virus (HIV) strains [Berger *et al.*, 1999]. Whereas G_i and $G\beta\gamma$ subunits are involved in CXCL12-induced Rac activation and cell migration of Jurkat T cells, $G\alpha_{13}$ mediates the activation of Rho by CXCR4 and that the functional activity of both $G\alpha_{13}$ and Rho is required for directional cell migration in response to CXCL12 [Tan *et al.*, 2006].

$S1P_3$ knockout mice are phenotypically normal, but depletion of $S1P_3$ abrogates a variety of S1P effects on the cardiovascular system [Ishii *et al.*, 2001]. S1P potently reduces myocardial perfusion in a manner dependent exclusively upon $S1P_3$ [Levkau *et al.*, 2004]. S1P induces chemotaxis of cells expressing $S1P_3$. However, upon inactivation of G_i by PTX treatment, $S1P/S1P_3$ couples to $G_{12/13}$ to mediate the inhibition of Rac and cell migration [Sugimoto *et al.*, 2003]. This result suggests a complex role of $S1P/S1P_3$ in the regulation of cell migration. Interception of the G_i-Ras-MEKK1 signaling pathway greatly inhibited LPA-stimulated migration of human ovarian cancer SK-OV3 cells, suggesting the involvement of G_i proteins [Bian *et al.*, 2004]. LPA-stimulated migration of human SK-OV3 cells is associated with $G_{12/13}$, but not G_q. LPA-induced FAK autophosphorylation and Rho activation are necessary for efficient LPA-stimulated cell migration [Bian *et al.*, 2006].

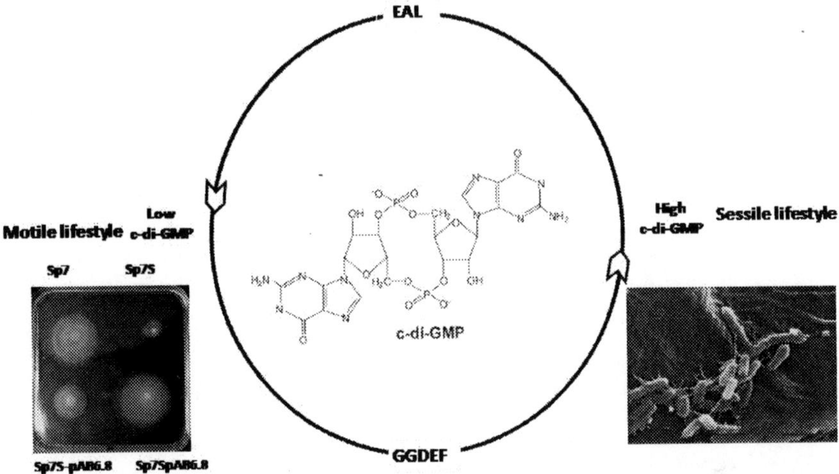

Figure 1. Integration of G protein signaling pathways in chemotaxis. The diagram illustrates the signaling molecules that participate in the signal integration of G_i and G_q (A), G_i and G_{12} (B), or G_q and G_{12} (C) that eventually lead to cell migration. Solid lined arrows indicate findings based on previous studies and dash lined arrows indicate putative interactions. The experimental evidence supporting individual pathways and the interactions between their intermediates are described in the text. CaM: calmodulin; PAK: p21-activated kinase; PI3K: phosphatidylinositol 3-kinase; PIP_3: phosphatidylinostiol(3,4,5)triphosphate; PIXα: PAK1-associated guanine nucleotide exchange factor; PTEN: 3'-phosphatase and tensin homolog; RhoGEF: Rho guanine nucleotide exchange factors; ROCK: Rho-dependent kinase; WASP: Wiskott-Aldrich syndrome protein.

Signal Integration of G_q And G_{12}

Crosstalk of the signaling pathways between G_q and $G_{12/13}$ has been reported in the studies of intercrosses of $G\alpha_q$-deficient mice and $G\alpha_{12}$-deficient mice [Gu *et al.*, 2002]. Both G_q and G_{12} proteins are capable of activating RhoA. $G\alpha_{12}$ and $G\alpha_{13}$ have been shown to activate RhoA by p115 family members, which consist of p115GEF, PDZ-RhoGEF and leukemia-associated RhoGEF (LARG). However, the direct interacting molecules linking G_q protein to RhoA remain elusive [Mizuno and Itoh, 2009]. $S1P_2$ is dispensable for murine cardiac development [Kono *et al.*, 2004]. S1P/$S1P_2$-mediated inhibitions of Rac and cell migration require both G_q and $G_{12/13}$ in rat VSMCs. S1P/$S1P_2$-mediated Rho activation in rat VSMCs is dependent on both G_q and $G_{12/13}$. The inhibition of Rac and cell migration is dependent on Rho activation [Takashima *et al.*, 2008].

LPC is a precursor of LPA and is generated by the conversion of phosphatidylcholine by PLA_2 [Maghazachi, 2005]. LPC/G2A-mediated chemotaxis of mouse primary macrophage and mouse macrophage J774A.1 cells are resistant to PTX treatment, while the C5a-induced cell migration was completely blocked [Yang *et al.*, 2005]. The expression of dominant negative forms of $G\alpha_q$, $G\alpha_{11}$, $G\alpha_{12}$ and $G\alpha_{13}$ suppressed the LPC/G2A-mediated chemotaxis of mouse macrophage J774A.1 cells, but not the C5a-stimulated migration. This shows the involvement of $G_{q/11}$ and $G_{12/13}$ proteins in LPC-mediated chemotaxis [Yang *et al.*, 2005].

Conclusion

Recent investigation has revealed an ever increasing evidence of previously unrecognized promiscuity of chemoattractant receptors in G protein coupling. Receptors for chemokine, S1P and LPA are coupled to G_i, G_q and $G_{12/13}$ protein families. However, which subfamilies of G protein are required in mediating the chemotactic response depends on their respective GPCR subtypes and the cell type under investigation. The dependence of GPCR subtypes can be illustrated in S1P-regulated chemotaxis. $S1P_1$ induces migratory response via G_i proteins whereas $S1P_2$-mediated inhibition of cell migration requires $G_{12/13}$ proteins. In addition, the usage of G protein subfamilies in cell mobilization is cell type specific. For example, CCL19 and CXCL12 stimulate migration of mouse T cells via G_i proteins whereas they activate chemotaxis of mouse dendritic cells through G_i and G_q proteins. A better understanding of their underlying molecular mechanisms in directed cell migration will help to generate more specific pharmacological tools for the treatment of pathologies caused by impaired cell migration. (4191 words)

Acknowledgments

Studies by the authors are supported in part by the Hong Kong Jockey Club, Research Grant Council (HKUST 644306, 643306 and 663108) and University Grant Council of Hong Kong (AoE/B-15/01) to YHW.

References

Adams, G. B., Alley, I. R., Chung, U. I., Chabner, K. T., Jeanson, N. T., Lo Celso, C., Marsters, E. S., Chen, M., Weinstein, L. S., Lin, C. P., Kronenberg, H. M. & Scadden, D. T. (2009). Haematopoietic stem cells depend on Gα_s-mediated signalling to engraft bone marrow. *Nature, 459(7243)*, 103-107.

Allavena, P., Bianchi, G., Zhou, D., van Damme, J., Jilek, P., Sozzani, S. & Mantovani, A. (1994). Induction of natural killer cell migration by monocyte chemotactic protein-1, -2 and -3. *Eur. J. Immunol., 24(12)*, 3233-3236.

Amatruda, T. T. 3rd, Gerard, N. P., Gerard, C. & Simon, M. I. (1993). Specific interactions of chemoattractant factor receptors with G-proteins. *J. Biol. Chem., 268(14)*, 10139-10144.

Amatruda, T. T. 3rd, Steele, D. A., Slepak, V. Z. & Simon, M. I. (1991). Gα_{16}, a G protein α subunit specifically expressed in hematopoietic cells. *Proc. Natl. Acad. Sci.,* U. S. A. *88(13)*, 5587-5591.

An, S., Bleu, T., Zheng, Y. & Goetzl, E. J. (1998). Recombinant human G protein-coupled lysophosphatidic acid receptors mediate intracellular calcium mobilization. *Mol. Pharmacol., 54(5)*, 881-888.

Andrews, S., Stephens, L. R. & Hawkins, P. T. (2007). PI3K class IB pathway in neutrophils. *Sci. STKE, 2007(407)*, cm3.

Arai, H., Tsou, C. L. & Charo, I. F. (1997). Chemotaxis in a lymphocyte cell line transfected with C-C chemokine receptor 2B: evidence that directed migration is mediated by $\beta\gamma$ dimers released by activation of Gα-coupled receptors. *Proc. Natl. Acad. Sci.,* U. S. A. *94(26)*, 14495-14499.

Berger, E. A., Murphy, P. M. & Farber, J. M. (1999). Chemokine receptors as HIV-1 coreceptors: roles in viral entry, tropism, and disease. *Annu. Rev. Immunol., 17*, 657-700.

Berridge, M. J. & Irvine, R. F. (1989). Inositol phosphates and cell signalling. *Nature, 341(6239)*, 197-205.

Bian, D., Mahanivong, C., Yu, J., Frisch, S. M., Pan, Z. K., Ye, R. D. & Huang, S. (2006). The $G_{12/13}$-RhoA signaling pathway contributes to efficient lysophosphatidic acid-stimulated cell migration. *Oncogene, 25(15)*, 2234-2244.

Bian, D., Su, S., Mahanivong, C., Cheng, R. K., Han, Q., Pan, Z. K., Sun, P. & Huang, S. (2004). Lysophosphatidic acid stimulates ovarian cancer cell migration via a Ras-MEK kinase 1 pathway. *Cancer Res., 64(12)*, 4209-4217.

Bourne, H. R. & Weiner, O. (2002). A chemical compass. *Nature, 419(6902)*, 21.

Bourne, H. R., Sanders, D. A. & McCormick, F. (1991). The GTPase superfamily: conserved structure and molecular mechanism. *Nature, 349(6305)*, 117-127.

Brown, S. L., Jala, V. R., Raghuwanshi, S. K., Nasser, M. W., Haribabu, B. & Richardson, R. M. (2006). Activation and regulation of platelet-activating factor receptor: role of G_i and G_q in receptor-mediated chemotactic, cytotoxic, and cross-regulatory signals. *J. Immunol., 177(5)*, 3242-3249.

Cambien, B., Pomeranz, M., Millet, M. A., Rossi, B. & Schmid-Alliana, A. (2001). Signal transduction involved in MCP-1-mediated monocytic transendothelial migration. *Blood, 97(2)*, 359-366.

Cassoni, P., Marrocco, T., Bussolati, B., Allia, E., Munaron, L., Sapino, A. & Bussolati, G. (2006). Oxytocin induces proliferation and migration in immortalized human dermal microvascular endothelial cells and human breast tumor-derived endothelial cells. *Mol. Cancer Res., 4(6)*, 351-359.

Cattaneo, M. G., Chini, B. & Vicentini, L. M. (2008). Oxytocin stimulates migration and invasion in human endothelial cells. *Br. J. Pharmacol., 153(4)*, 728-736.

Chaulet, H., Desgranges, C., Renault, M. A., Dupuch, F., Ezan, G., Peiretti, F., Loirand, G., Pacaud, P. & Gadeau, A. P. (2001). Extracellular nucleotides induce arterial smooth muscle cell migration via osteopontin. *Circ. Res., 89(9)*, 772-778.

Chen, Y., Corriden, R., Inoue, Y., Yip, L., Hashiguchi, N., Zinkernagel, A., Nizet, V., Insel, P. A. & Junger, W. G. (2006). ATP release guides neutrophil chemotaxis via $P2Y_2$ and A_3 receptors. *Science, 314(5806)*, 1792-1795.

Dhanasekaran, N. & Dermott, J. M. (1996). Signaling by the G_{12} class of G proteins. *Cell. Signal., 8(4)*, 235-245.

Donnini, S., Finetti, F., Solito, R., Terzuoli, E., Sacchetti, A., Morbidelli, L., Patrignani, P. & Ziche, M. (2007). EP2 prostanoid receptor promotes squamous cell carcinoma growth through epidermal growth factor receptor transactivation and iNOS and ERK1/2 pathways. *FASEB J., 21(10)*, 2418-2430.

Frade, J. M., Mellado, M., del Real, G., Gutierrez-Ramos, J. C., Lind, P. & Martinez-A, C. (1997). Characterization of the CCR2 chemokine receptor:

functional CCR2 receptor expression in B cells. *J. Immunol., 159(11),* 5576-5584.

Fukuhara, S., Chikumi, H. & Gutkind, J. S. (2000). Leukemia-associated Rho guanine nucleotide exchange factor (LARG). links heterotrimeric G proteins of the G_{12} family to Rho. *FEBS Lett., 485(2-3)*, 183-188

Fukuhara, S., Murga, C., Zohar, M., Igishi, T. & Gutkind JS (1999). A novel PDZ domain containing guanine nucleotide exchange factor links heterotrimeric G proteins to Rho. *J. Biol. Chem., 274(9),* 5868-5879.

Gu, J. L., Müller, S., Mancino, V., Offermanns, S. & Simon, M. I. (2002). Interaction of $G\alpha_{12}$ with $G\alpha_{13}$ and $G\alpha_q$ signaling pathways. *Proc. Natl. Acad. Sci.,* U. S. A. *99(14),* 9352-9357.

Hall, A. (1998). Rho GTPases and the actin cytoskeleton. *Science, 279(5350),* 509-514.

Hall, D. A., Beresford, I. J., Browning, C. & Giles, H. (1999). Signalling by CXC-chemokine receptors 1 and 2 expressed in CHO cells: a comparison of calcium mobilization, inhibition of adenylyl cyclase and stimulation of GTPγS binding induced by IL-8 and GROα. *Br. J. Pharmacol., 126(3),* 810-818.

Haribabu, B., Zhelev, D. V., Pridgen, B. C., Richardson, R. M., Ali, H. & Snyderman R. (1999). Chemoattractant receptors activate distinct pathways for chemotaxis and secretion. Role of G-protein usage. *J. Biol. Chem., 274(52),* 37087-37092.

Hart, M. J., Jiang, X., Kozasa, T., Roscoe, W., Singer, W. D., Gilman, A. G., Sternweis, P. C. & Bollag, G. (1998). Direct stimulation of the guanine nucleotide exchange activity of p115 RhoGEF by $G\alpha_{13}$. *Science, 280(5372),* 2112-2114.

Hirshman, C. A. & Emala, C. W. (1999). Actin reorganization in airway smooth muscle cells involves G_q and G_{i2} activation of Rho. *Am. J. Physiol., 277(3 Pt 1),* L653-661.

Ho, M. K., Yung, L. Y., Chan, J. S., Chan, J. H., Wong, C. S. & Wong, Y. H. (2001). $G\alpha_{14}$ links a variety of G_i- and G_s-coupled receptors to the stimulation of phospholipase C. *Br. J. Pharmacol., 132(7),* 1431-1440.

Hooley, R., Yu, C. Y., Symons, M. & Barber, D. L. (1996). $G\alpha_{13}$ stimulates Na^+-H^+ exchange through distinct Cdc42-dependent and RhoA-dependent pathways. *J. Biol. Chem., 271(11),* 6152-6158.

Hwang, J. I., Fraser, I. D., Choi, S., Qin, X. F. & Simon, M. I. (2004). Analysis of C5a-mediated chemotaxis by lentiviral delivery of small interfering RNA. *Proc. Natl. Acad. Sci.,* U. S. A. *101(2),* 488-493.

Ishii, I., Friedman, B., Ye, X., Kawamura, S., McGiffert, C., Contos, J. J., Kingsbury, M. A., Zhang, G., Brown, J. H. & Chun, J. (2001). Selective loss of sphingosine 1-phosphate signaling with no obvious phenotypic abnormality in mice lacking its G protein-coupled receptor, LP_{B3}/EDG-3. *J. Biol. Chem., 276(36),* 33697-33704.

Jeong, K. J., Park, S. Y., Seo, J. H., Lee, K. B., Choi, W. S., Han, J. W., Kang, J. K., Park, C. G., Kim, Y. K. & Lee, H. Y. (2008). Lysophosphatidic acid receptor 2 and G_i/Src pathway mediate cell motility through cyclooxygenase 2 expression in CAOV-3 ovarian cancer cells. *Exp. Mol. Med. 40(6),* 607-616.

Jiang, H., Wu, D. & Simon, M. I. (1994). Activation of phospholipase C β4 by heterotrimeric GTP-binding proteins. *J. Biol. Chem. 269(10),* 7593-7596.

Juarez, J. & Bendall, L. (2004). SDF-1 and CXCR4 in normal and malignant hematopoiesis. *Histol. Histopathol., 19(1),* 299-309.

Kim, J., Keys, J. R. & Eckhart, A. D. (2006). Vascular smooth muscle migration and proliferation in response to lysophosphatidic acid (LPA). is mediated by LPA receptors coupling to G_q. *Cell. Signal., 18(10),* 1695-1701.

Kim, Y., Bae, Y. S., Park, J. C., Suh, P. G. & Ryu, S. H. (2001). The synthetic peptide, His-Phe-Tyr-Leu-Pro-Met, is a chemoattractant for Jukat T cells. *Exp. Mol. Med., 33(4),* 257-262.

Ko, J., Kim, I. S., Jang, S. W., Lee, Y. H., Shin, S. Y., Min do, S. & Na, D. S. (2002). Leukotactin-1/CCL15-induced chemotaxis signaling through CCR1 in HOS cells. *FEBS Lett., 515(1-3),* 159-164.

Koizumi, K., Hojo, S., Akashi, T., Yasumoto, K. & Saiki, I. (2007). Chemokine receptors in cancer metastasis and cancer cell-derived chemokines in host immune response. *Cancer Sci., 98(11),* 1652-1658.

Kono, M., Mi, Y., Liu, Y., Sasaki, T., Allende, M. L., Wu, Y. P., Yamashita, T. & Proia, R. L. (2004). The sphingosine-1-phosphate receptors $S1P_1$, $S1P_2$, and $S1P_3$ function coordinately during embryonic angiogenesis. *J. Biol. Chem., 279(28),* 29367-29373.

Kowalska, M. A., Ratajczak, M. Z., Majka, M., Jin, J., Kunapuli, S., Brass, L. & Poncz, M. (2000). Stromal cell-derived factor-1 and macrophage-derived chemokine: 2 chemokines that activate platelets. *Blood, 96(1),* 50-57.

Kozasa, T., Jiang, X., Hart, M. J., Sternweis, P. M., Singer, W. D., Gilman, A. G., Bollag, G. & Sternweis, P. C. (1998). p115 RhoGEF, a GTPase activating protein for $G\alpha_{12}$ and $G\alpha_{13}$. *Science, 280(5372),* 2109-2111.

Kranenburg, O., Poland, M., van Horck, F. P., Drechsel, D., Hall, A. & Moolenaar, W. H. (1999). Activation of RhoA by lysophosphatidic acid and $G\alpha_{12/13}$ subunits in neuronal cells: induction of neurite retraction. *Mol. Biol. Cell, 10(6),* 1851-1857.

Kuang, Y., Wu, Y., Jiang, H. & Wu, D. (1996). Selective G protein coupling by C-C chemokine receptors. *J. Biol. Chem., 271(8),* 3975-3978.

Lee, C. H., Park, D., Wu, D., Rhee, S. G. & Simon, M. I. (1992). Members of the $G_q\alpha$ subunit gene family activate phospholipase Cβ isozymes. *J. Biol. Chem., 267(23),* 16044-16047.

Lee, J. W., Joshi, S., Chan, J. S. & Wong, Y. H. (1998). Differential coupling of μ-, δ-, and κ-opioid receptors to $G\alpha_{16}$-mediated stimulation of phospholipase C. *J. Neurochem., 70(5),* 2203-2211.

Lee, O. H., Kim, Y. M., Lee, Y. M., Moon, E. J., Lee, D. J., Kim, J. H., Kim, K. W. & Kwon, Y. G. (1999). Sphingosine 1-phosphate induces angiogenesis: its angiogenic action and signaling mechanism in human umbilical vein endothelial cells. *Biochem. Biophys. Res. Commun., 264(3),* 743-750.

Levkau, B., Hermann, S., Theilmeier, G., van der Giet, M., Chun, J., Schober, O. & Schäfers, M. (2004). High-density lipoprotein stimulates myocardial perfusion *in vivo*. *Circulation, 110(21),* 3355-3359.

Li, Z., Hannigan, M., Mo, Z., Liu, B., Lu, W., Wu, Y., Smrcka, A. V., Wu, G., Li, L., Liu, M., Huang, C. K. & Wu, D. (2003). Directional sensing requires Gβγ-mediated PAK1 and PIXα-dependent activation of Cdc42. *Cell, 114(2),* 215-227.

Liehn, E. A., Zernecke, A., Postea, O. & Weber, C. (2006). Chemokines: inflammatory mediators of atherosclerosis. *Arch. Physiol. Biochem., 112(4-5),* 229-238.

Lin, X., Voyno-Yasenetskaya, T. A., Hooley, R., Lin, C. Y., Orlowski, J. & Barber, D. L. (1996). $G\alpha_{12}$ differentially regulates Na^+-H^+ exchanger isoforms. *J. Biol. Chem., 271(37),* 22604-22610.

Liu, Y., Wada, R., Yamashita, T., Mi, Y., Deng, C. X., Hobson, J. P., Rosenfeldt, H. M., Nava, V. E., Chae, S. S., Lee, M. J., Liu, C. H., Hla, T., Spiegel, S. & Proia, R. L. (2000). Edg-1, the G protein-coupled receptor for sphingosine-1-phosphate, is essential for vascular maturation. *J. Clin. Invest., 106(8),* 951-961.

Maghazachi, A. A. (2005). Insights into seven and single transmembrane-spanning domain receptors and their signaling pathways in human natural killer cells. *Pharmacol. Rev., 57(3),* 339-357.

Maghazachi, A. A., Al-Aoukaty, A. & Schall, T. J. (1994). C-C chemokines induce the chemotaxis of NK and IL-2-activated NK cells. Role for G proteins. *J. Immunol., 153(11),* 4969-4977.

Matloubian, M , Lo, C. G., Cinamon, G., Lesneski, M. J., Xu, Y., Brinkmann, V., Allende, M. L., Proia, R. L. & Cyster, J. G. (2004). Lymphocyte egress from thymus and peripheral lymphoid organs is dependent on S1P receptor 1. *Nature, 427(6972),* 355-360.

McLaughlin, S. K., McKinnon, P. J., Spickofsky, N., Danho, W. & Margolskee, R. F. (1994). Molecular cloning of G proteins and phosphodiesterases from rat taste cells. *Physiol. Behav. 56(6),* 1157-1164.

Mills, G. B. & Moolenaar, W. H.(2003). The emerging role of lysophosphatidic acid in cancer. *Nat. Rev. Cancer, 3(8),* 582-591.

Mizuno, N. & Itoh, H. (2009). Functions and regulatory mechanisms of G_q-signaling pathways. *Neurosignals, 17(1),* 42-54.

Morita, T., Shibata, K., Kikkawa, F., Kajiyama, H., Ino, K. & Mizutani, S. (2004). Oxytocin inhibits the progression of human ovarian carcinoma cells *in vitro* and *in vivo*. *Int. J. Cancer, 109(4),* 525-532.

Myers, S. J., Wong, L. M. & Charo, I. F. (1995). Signal transduction and ligand specificity of the human monocyte chemoattractant protein-1 receptor in transfected embryonic kidney cells. *J. Biol. Chem., 270(11),* 5786-5792.

Nakamura, F., Ogata, K., Shiozaki, K., Kameyama, K., Ohara, K., Haga, T. & Nukada, T. (1991). Identification of two novel GTP-binding protein α-subunits that lack apparent ADP-ribosylation sites for pertussis toxin. *J. Biol. Chem., 266(19),* 12676-12681.

Neer, E. J. (1995). Heterotrimeric G proteins: organizers of transmembrane signals. *Cell, 80(2),* 249-257.

Neptune, E. R. & Bourne, H. R. (1997). Receptors induce chemotaxis by releasing the βγ subunit of G_i, not by activating G_q or G_s. *Proc. Natl. Acad. Sci., U. S .A. 94(26),* 14489-14494.

Offermanns, S. & Simon, M. I. (1995). $G\alpha_{15}$ and $G\alpha_{16}$ couple a wide variety of receptors to phospholipase C. *J. Biol. Chem., 270(25),* 15175-15180.

Offermanns, S., Mancino, V., Revel, J. P. & Simon, M. I. (1997). Vascular system defects and impaired cell chemokinesis as a result of $G\alpha_{13}$ deficiency. *Science, 275(5299),* 533-536.

Okamoto, H., Takuwa, N., Yokomizo, T., Sugimoto, N., Sakurada, S., Shigematsu, H. & Takuwa, Y. (2000). Inhibitory regulation of Rac activation, membrane ruffling, and cell migration by the G protein-coupled sphingosine-1-phosphate receptor EDG5 but not EDG1 or EDG3. *Mol. Cell. Biol., 20(24),* 9247-9261.

Olson, T. S. & Ley, K. (2002). Chemokines and chemokine receptors in leukocyte trafficking. *Am. J. Physiol. Regul. Integr. Comp. Physiol., 283(1),* R7-R28.

Ono, S. J., Nakamura, T., Miyazaki, D., Ohbayashi, M., Dawson, M. & Toda, M. (2003). Chemokines: roles in leukocyte development, trafficking, and effector function. *J. Allergy Clin. Immunol., 111(6),* 1185-1199.

Pillois, X., Chaulet, H., Belloc, I., Dupuch, F., Desgranges, C. & Gadeau, A. P. (2002). Nucleotide receptors involved in UTP-induced rat arterial smooth muscle cell migration. *Circ. Res., 90(6),* 678-681.

Pintor, J., Bautista, A., Carracedo, G. & Peral, A. (2004). UTP and diadenosine tetraphosphate accelerate wound healing in the rabbit cornea. *Ophthalmic Physiol. Opt., 24(3),* 186-193.

Ponath, P. D., Qin, S., Post, T. W., Wang, J., Wu, L., Gerard, N. P., Newman, W., Gerard, C. & Mackay, C. R. (1996). Molecular cloning and characterization of a human eotaxin receptor expressed selectively on eosinophils. *J. Exp. Med., 183(6),* 2437-2448.

Radeff-Huang, J., Seasholtz, T. M., Matteo, R. G. & Brown, J. H. (2004). G protein mediated signaling pathways in lysophospholipid induced cell proliferation and survival. *J. Cell. Biochem., 92(5),* 949-966.

Ridley, A. J. (1997). The GTP-binding protein Rho. *Int. J. Biochem. Cell. Biol., 29(11),* 1225-1229.

Sadana, R. & Dessauer, C. W. (2009). Physiological roles for G protein-regulated adenylyl cyclase isoforms: insights from knockout and overexpression studies. *Neurosignals, 17(1),* 5-22.

Sebok, K., Woodside, D., al-Aoukaty, A., Ho, A. D., Gluck, S. & Maghazachi, A. A. (1993). IL-8 induces the locomotion of human IL-2-activated natural killer cells. Involvement of a guanine nucleotide binding (G_o). protein. *J. Immunol., 150(4),* 1524-1534.

Shi, G., Partida-Sánchez, S., Misra, R. S., Tighe, M., Borchers, M. T., Lee, J. J., Simon, M. I. & Lund, F. E. (2007). Identification of an alternative $G\alpha_q$-dependent chemokine receptor signal transduction pathway in dendritic cells and granulocytes. *J. Exp. Med., 204(11),* 2705-2718.

Shyamala, V. & Khoja, H. (1998). Interleukin-8 receptors R1 and R2 activate mitogen-activated protein kinases and induce c-fos, independent of Ras and Raf-1 in Chinese hamster ovary cells. *Biochemistry, 37(45),* 15918-15924.

Siess, W. (2002). Athero- and thrombogenic actions of lysophosphatidic acid and sphingosine-1-phosphate. *Biochim. Biophys. Acta, 1582(1-3),* 204-215.

Silva, T. A., Garlet, G. P., Fukada, S. Y., Silva, J. S. & Cunha, F. Q. (2007). Chemokines in oral inflammatory diseases: apical periodontitis and periodontal disease. *J. Dent. Res., 86(4),* 306-319.

Simon, M. I., Strathmann, M. P. & Gautam, N. (1991). Diversity of G proteins in signal transduction. *Science, 252(5007),* 802-808.

Soriano, S. F., Serrano, A., Hernanz-Falcon, P., Martin de Ana, A., Monterrubio, M., Martinez, C., Rodriguez-Frade, J. M. & Mellado, M. (2003). Chemokines integrate JAK/STAT and G-protein pathways during chemotaxis and calcium flux responses. *Eur. J. Immunol., 33(5),* 1328-1333.

Sotsios, Y., Whittaker, G. C., Westwick, J. & Ward, S. G. (1999). The CXC chemokine stromal cell-derived factor activates a G_i-coupled phosphoinositide 3-kinase in T lymphocytes. *J. Immunol., 163(11)*, 5954-5963.

Sozzani, S., Luini, W., Molino, M., Jilek, P., Bottazzi, B., Cerletti, C., Matsushima, K. & Mantovani, A. (1991). The signal transduction pathway involved in the migration induced by a monocyte chemotactic cytokine. *J. Immunol., 147(7),* 2215-2221.

Sozzani, S., Zhou, D., Locati, M., Rieppi, M., Proost, P., Magazin, M., Vita, N., van Damme, J. & Mantovani, A. (1994). Receptors and transduction pathways for monocyte chemotactic protein-2 and monocyte chemotactic protein-3. Similarities and differences with MCP-1. *J. Immunol., 152(7),* 3615-3622.

Spiegel, A. M. (1992). G proteins in cellular control. *Curr. Opin. Cell. Biol., 4(2),* 203-211.

Stradal, T. E., Rottner, K., Disanza, A., Confalonieri, S., Innocenti, M. & Scita, G. (2004). Regulation of actin dynamics by WASP and WAVE family proteins. *Trends Cell. Biol., 14(6),* 303-311.

Su, Y., Ho, M. K. & Wong, Y. H. (2009). A hematopoietic perspective on the promiscuity and specificity of $G\alpha_{16}$ signaling. *Neurosignals, 17(1),* 71-81.

Sugimoto, N., Takuwa, N., Okamoto, H., Sakurada, S. & Takuwa, Y. (2003). Inhibitory and stimulatory regulation of Rac and cell motility by the $G_{12/13}$-Rho and G_i pathways integrated downstream of a single G protein-coupled sphingosine-1-phosphate receptor isoform. *Mol. Cell. Biol., 23(5),* 1534-1545.

Suzuki, N., Hajicek, N. & Kozasa, T. (2009). Regulation and physiological functions of $G_{12/13}$-mediated signaling pathways. *Neurosignals, 17(1),* 55-70.

Taboubi, S., Milanini, J., Delamarre, E., Parat, F., Garrouste, F., Pommier, G., Takasaki, J., Hubaud, J. C., Kovacic, H. & Lehmann, M. (2007). $G\alpha_{q/11}$-coupled $P2Y_2$ nucleotide receptor inhibits human keratinocyte spreading and migration. *FASEB J., 21(14),* 4047-4058.

Taha, T. A., Argraves, K. M. & Obeid, L. M. (2004). Sphingosine-1-phosphate receptors: receptor specificity versus functional redundancy. *Biochim. Biophys. Acta., 1682(1-3),* 48-55.

Takashima, S., Sugimoto, N., Takuwa, N., Okamoto, Y., Yoshioka, K., Takamura, M., Takata, S., Kaneko, S. & Takuwa, Y. (2008). $G_{12/13}$ and G_q mediate $S1P_2$-induced inhibition of Rac and migration in vascular smooth muscle in a manner dependent on Rho but not Rho kinase. *Cardiovasc. Res., 79(4),* 689-697.

Tan, W., Martin, D. & Gutkind, J. S. (2006). The $G\alpha_{13}$-Rho signaling axis is required for SDF-1-induced migration through CXCR4. *J. Biol. Chem., 281(51),* 39542-39549.

Thompson, B. D., Jin, Y., Wu, K. H., Colvin, R. A., Luster, A. D., Birnbaumer, L. & Wu, M. X. (2007). Inhibition of $G\alpha_{i2}$ activation by $G\alpha_{i3}$ in CXCR3-mediated signaling. *J. Biol. Chem., 282(13),* 9547-9555.

Tian, Y. J., Lee, M. M., Yung, L. Y., Allen, R. A., Slocombe, P. M., Twomey, B. M. & Wong, Y. H. (2008). Differential involvement of $G\alpha_{16}$ in CC chemokine-induced stimulations of phospholipase Cβ, ERK, and chemotaxis. *Cell. Signal., 20*, 1179-1189.

Vila-Coro, A. J., Rodríguez-Frade, J. M., Martín De Ana, A., Moreno-Ortíz, M. C., Martínez-A, C. & Mellado, M. (1999). The chemokine SDF-1α triggers CXCR4 receptor dimerization and activates the JAK/STAT pathway. *FASEB J., 13(13),* 1699-1710.

Walsh, M. P., Andrea, J. E., Allen, B. G., Clement-Chomienne, O., Collins, E. M. & Morgan, K. G. (1994). Smooth muscle protein kinase C. *Can. J. Physiol. Pharmacol., 72(11),* 1392-1399.

Wang, F., Van Brocklyn, J. R., Hobson, J. P., Movafagh, S., Zukowska-Grojec, Z., Milstien, S. & Spiegel, S. (1999). Sphingosine 1-phosphate stimulates cell migration through a G_i-coupled cell surface receptor. Potential involvement in angiogenesis. *J. Biol. Chem., 274(50),* 35343-35350.

Wang, M., Kong, Q., Gonzalez, F. A., Sun, G., Erb, L., Seye, C. & Weisman, G. A. (2005). P2Y nucleotide receptor interaction with α integrin mediates astrocyte migration. *J. Neurochem., 95(3),* 630-640.

Weiner, O. D. (2002). Rac activation: P-Rex1 - a convergence point for PIP_3 and G$\beta\gamma$? *Curr. Biol., 12(12),* R429-R431.

Welch, H. C., Coadwell, W. J., Ellson, C. D., Ferguson, G. J., Andrews, S. R., Erdjument-Bromage, H., Tempst, P., Hawkins, P. T. & Stephens, L. R. (2002). P-Rex1, a PtdIns(3,4,5).P_3- and G$\beta\gamma$-regulated guanine-nucleotide exchange factor for Rac., *Cell 108(6),* 809-821.

Wilkie, T. M., Scherle, P. A., Strathmann, M. P., Slepak, V. Z. & Simon, M. I. (1991). Characterization of G-protein α subunits in the G_q class: expression in murine tissues and in stromal and hematopoietic cell lines. *Proc. Natl. Acad. Sci.,* U. S. A. *88(22),* 10049-10053.

Wong, K., Van Keymeulen, A. & Bourne, H. R. (2007). PDZRhoGEF and myosin II localize RhoA activity to the back of polarizing neutrophil-like cells. *J. Cell. Biol., 179(6),* 1141-1148.

Wu, D., LaRosa, G. J. & Simon, M. I. (1993). G protein-coupled signal transduction pathways for interleukin-8. *Science, 261(5117),* 101-103.

Xu, J., Wang, F., Van Keymeulen, A., Herzmark, P., Straight, A., Kelly, K., Takuwa, Y., Sugimoto, N., Mitchison, T. & Bourne, H. R. (2003). Divergent signals and cytoskeletal assemblies regulate self-organizing polarity in neutrophils., *Cell 114(2),* 201-214.

Yang, L. V., Radu, C. G., Wang, L., Riedinger, M. & Witte, O. N. (2005). G_i-independent macrophage chemotaxis to lysophosphatidylcholine via the immunoregulatory GPCR G2A. *Blood, 105(3),* 1127-1134.

Yang, L., Huang, Y., Porta, R., Yanagisawa, K., Gonzalez, A., Segi, E., Johnson, D. H., Narumiya, S. & Carbone, D. P. (2006). Host and direct antitumor effects and profound reduction in tumor metastasis with selective EP4 receptor antagonism. *Cancer Res., 66(19)*, 9665-9672.

Ye, X., Ishii, I., Kingsbury, M. A. & Chun, J. (2002) Lysophosphatidic acid as a novel cell survival/apoptotic factor. *Biochim. Biophys. Acta., 1585(2-3),* 108-13.

Yen, H., Zhang, Y., Penfold, S. & Rollins, B. J. (1997). MCP-1-mediated chemotaxis requires activation of non-overlapping signal transduction pathways. *Leukoc. Biol., 61(4)*, 529-532.

Yoon, C. M., Hong, B. S., Moon, H. G., Lim, S., Suh, P. G., Kim, Y. K., Chae, C. B. & Gho, Y. S. (2008). Sphingosine-1-phosphate promotes lymphangiogenesis by stimulating $S1P_1/G_i/PLC/Ca^{2+}$ signaling pathways. *Blood, 112(4)*, 1129-1138.

Zhao, J., Ma, L., Wu, Y. L., Wang, P., Hu, W. & Pei, G. (1998). Chemokine receptor CCR5 functionally couples to inhibitory G proteins and undergoes desensitization. *J. Cell. Biochem. 71(1)*, 36-45.

Zigman, J. M., Westermark, G. T., LaMendola, J. & Steiner, D. F. (1994). Expression of cone transducin, $G_z\alpha$, and other G-protein α-subunit messenger ribonucleic acids in pancreatic islets. *Endocrinology, 135(1)*, 31-37.

In: Chemotaxis: Types, Clinical Significance… ISBN: 978-1-61728-495-3
Editor: T. C. Williams, pp. 135-156

Chapter 5

Chemotaxis in the Model Organism *Chlamydomonas Reinhardtii*

***E.G. Govorunova*[1*], *O.O. Voytsekh*[1,3], *A.P. Filonova*[1,3], *M.A. Kutuzov*[2**], *M. Mittag*[3] and *O.A. Sineshchekov*[1]**

[1]Biology Department, M.V. Lomonosov Moscow State University, Vorobievy Gory, 119992 Moscow, Russia

[2]Scottish Crop Research Institute, Invergowrie, Dundee DD2 5DA Scotland, UK

[3]Institut für Allgemeine Botanik und Pflanzenphysiologie, Friedrich-Schiller-Universität Jena, 07743 Jena, Germany

Abstract

The unicellular biflagellate alga *Chlamydomonas reinhardtii* is widely used as a simple model to study fundamental cellular processes, including chemotaxis. *C. reinhardtii* is attracted towards ammonium ions (NH_4^+) and peptide mixtures, such as a pancreatic digest of casein (tryptone). The sensitivity to NH_4^+ is transiently induced in vegetative

* Corresponding author, e-mail: egovoru@yahoo.com

** Present address: Department of Pharmacology, University of Illinois at Chicago, Chicago, IL 60612, USA cells by nitrogen deprivation, whereas the sensitivity to tryptone requires formation of mature gametes. The clock-controlled RNA-binding protein CHLAMY1 might be involved in regulation of the diurnal rhythm of chemotaxis to NH_4^+. We measured inhibition of rhodopsin-mediated photoreceptor currents by the chemoattractant tryptone as an indirect assay of early stages of chemosensory transduction in *C. reinhardtii* gametes. The results showed that the magnitude of the response to tryptone depended on the concentration of monovalent metal cations in the medium with the selectivity sequence $K^+>Rb^+>Cs^+>Na^+>Li^+$. It was inhibited by extracellular Ba^{2+} and Ca^{2+} in millimolar concentrations, and by dibutyryl-cAMP. These observations suggest the involvement of K^+ channels modulated by cyclic nucleotides in *C. reinhardtii* chemotaxis. In order to identify active ingredients, we subjected tryptone to fractionation by gel filtration and demonstrated that only fractions that contain individual amino acids and dipeptides were functionally active. However, none of the 18 tested individual amino acids fully mimicked the effect of tryptone. We hypothesize that a specific combination of amino acids and/or dipeptides acts as a chemoattractant for *C. reinhardtii* gametes.

Introduction

Chlamydomonas reinhardtii, a motile photosynthetic alga, is a preferred unicellular model to study many biochemical and physiological processes. Easy cultivation, cell fractionation and manipulation of genetic material are several of many advantages of this microorganism for research. Its nuclear, mitochondrial and chloroplast genomes are completely sequenced. The behavior of *C. reinhardtii* is controlled by environmental stimuli, including light and chemical cues. Mechanisms of light sensing in *C. reinhardtii* and related flagellates are relatively well characterized (for review see Sineshchekov and Spudich, 2005; Hegemann and Berthold, 2009; Kreimer, 2009). Two species of sensory rhodopsins (also known as channelrhodopsins) serve as photoreceptors for motility responses in *C. reinhardtii* (Sineshchekov et al., 2002; Nagel et al., 2002, 2003; Govorunova et al., 2004). Photoexcitation of either rhodopsin results in generation of inward photoreceptor currents (PCs) across the plasma membrane, which initiates a signaling cascades leading to phototaxis and photoshock responses (Litvin et al., 1978; Harz and Hegemann, 1991). PCs elicited by short flashes of light can be recorded from freely swimming *C. reinhardtii* cells with a technically simple population method (Sineshchekov et al., 1992).

The chemobehavior in *C. reinhardtii* has been mainly studied with capillary assays, i.e., by measuring accumulation of cells inside or in front of a glass capillary filled with a solution of a chemoattractant. Ammonium ions (NH_4^+) and protein hydrolysates, such as tryptone, have been identified by this technique as two strongest chemoattractants for *C. reinhardtii* (Sjoblad and Frederikse, 1981; Govorunova and Sineshchekov, 2003). However, cellular and molecular mechanisms of chemosensory transduction in *C. reinhardtii* remain almost fully obscure. In this context, it is worthwhile mentioning that some putative chemotaxis-related proteins have been identified in the proteome of the eyespot apparatus (Schmidt et al., 2006). It seems plausible that the chemosensory signaling mechanisms involve changes in the electrical conductance of the cell membrane. But, so far no experimental techniques have been developed to verify this hypothesis directly. Therefore, we have developed an indirect approach to probe chemosensing in *C. reinhardtii* by monitoring changes in PCs produced by the addition of the chemoattractant tryptone to *C. reinhardtii* suspensions (Govorunova and Sineshchekov, 2003). We found that tryptone strongly inhibited PCs in mature gametes, but not in vegetative cells. On the other hand, we observed that only gametes, but not vegetative cells, exhibited chemoaccumulation when exposed to tryptone. Moreover, the magnitude of the inhibition of PCs correlated with the strength of chemoaccumulation of the cells upon variation of experimental conditions. This correlation suggests that the tryptone-induced inhibition of PCs reflects activation of a gamete-specific sensory system that mediates chemotaxis to tryptone. The addition of NH_4^+ to *C. reinhardtii* cell suspensions also caused inhibition of PCs. However, no correlation with chemoaccumulation was observed in this case (Govorunova and Sineshchekov, 2005). Therefore, chemotactic responses to tryptone and NH_4^+ appear to involve different signaling pathways.

Mature *C. reinhardtii* gametes do not respond to NH_4^+ (Byrne et al., 1992; Ermilova et al., 2003). In vegetative cells, nitrogen deprivation induces both chemotaxis to this ion (Sjoblad and Frederikse, 1981) and gametogenesis (Sager and Granick, 1954). It was however unclear, whether there is a relationship between these two processes. In this study, we demonstrate that chemotaxis to NH_4^+ can be induced independently of gametogenesis, which also distinguishes it from chemotaxis to tryptone.

Chemotaxis to NH_4^+ is strongly regulated by the circadian clock, with the peak in the middle of the subjective night (Byrne et al., 1992). Circadian rhythms are endogenous biological programs that provide coordination of biochemical, physiological and behavioral processes with the daily cycle (for review see Johnson, 2001). *C. reinhardtii* is extensively used to analyze molecular mechanisms of the circadian clock by biochemical, genetic and bioinformatics approaches (for review see Mittag and Wagner, 2003; Schulze et al., 2010). Several essential elements of the circadian clock known in other well-characterized organisms are conserved in *C. reinhardtii*, but others appear to be unique for this protist (Mittag et al., 2005). In the past years, several components of the oscillatory machinery of *C. reinhardtii* have been characterized (Iliev et al., 2006; Schmidt et al., 2006; Matsuo et al., 2008; Serrano et al., 2009). Based on functional homology with the circadian controlled translational regulator (CCTR) from the dinoflagellate *Gonyaulax polyedra*, an RNA-binding protein had been identified in *C. reinhardtii* and was named CHLAMY1 (Mittag, 1996). This heteromeric protein consists of two subunits, named C1 and C3 (Zhao et al., 2004). Silencing or overexpression of the genes encoding either of these subunits disturbs the circadian rhythms of phototaxis and nitrite reductase activity (Iliev et al., 2006). Here we show that the C3 subunit may also be involved in regulation of the chemotaxis diurnal rhythm.

Measurement of the inhibition of PCs by tryptone enables probing chemosensory transduction with an improved time resolution and specificity, as compared to measurement of chemoaccumulation. In this study, we used this approach to further characterize gamete-specific chemosensory transduction in *C. reinhardtii*. We examined the dependence of the response to tryptone on the extracellular concentration of monovalent metal cations, K^+ channel blockers and a permeable analog of cAMP. The results suggest that cAMP-gated K^+ channels might mediate chemosensing in *C. reinhardtii* gametes. Finally, we tested the response to different tryptone fractions and individual amino acids in order to characterize chemoattractant substances, and found them to be small molecules, possibly dipeptides, or amino acids acting in concert.

Materials and Methods

Strains and Culture Conditions

The photosynthetic flagellate *Chlamydomonas reinhardtii* Dangeard strain 495 *mt+* was provided by Dr. A.S. Chunaev (St. Petersburg University, St. Petersburg, Russia). Strain C3-sil$_6$ with a reduced content of the CHLAMY1

C3 subunit was generated as reported earlier (Iliev et al., 2006) in SAG 73.72 genetic background. Cells were grown in liquid high salt acetate medium (HSA; Harris, 1989) under continuous illumination (10 W m^{-2}) from cool fluorescence lamps at 25°C, except experiments in which the diurnal rhythm of chemotaxis was measured. In this case, cells were grown under an LD 12:12 cycle for 3-4 cycles to entrain the circadian clock. Aliquots of culture were collected during the next 2 cycles to measure chemotaxis, and the data obtained during the 1st and 2nd cycles were pooled together. Each aliquot was transferred to nitrogen-deficient minimal medium (NMM: 3.1 mM K_2HPO_4, 3.4 mM KH_2PO_4, 81 μM $MgSO_4$, 100 μM $CaCl_2$; Harris, 1989) and kept for 1 h at the density 10^7 cells x mL^{-1} on a rotary shaker under continuous illumination (2 W m^{-2}) to induce chemotactic sensitivity to NH_4^+. Mature gametes were obtained by overnight incubation under the same conditions (Beck and Haring, 1996). Nitrogen-deprived vegetative cells were prepared by incubation in NMM in the dark. Quantification of the fraction of mature (mating-competent) gametes in cell populations was performed as described earlier (Govorunova et al., 2007). Bacto tryptone was from Difco Laboratories (Detroit, MI), db-cAMP (sodium salt) from ICN (Irvine, CA), 4-aminopyridine from Sigma-Aldrich (St. Louis, MO). All other chemicals were of analytical grade.

Measurement of Chemoaccumulation

Chemoaccumulation of *C. reinhardtii* cells in front of the capillary filled with a chemottractant solution was measured as described in Govorunova et al. (2007). The optical density was recorded with a photodiode (K-24, Russia), the output of which was fed to a current amplifier (model 428, Keithley Instruments, Cleveland, OH). A monitoring light beam was provided by a microscope illuminator supplemented with a cut-off filter (> 730 nm) to avoid photoaccumulation of the cells. The amplifier output signals were digitized by a MiniDigi interface controlled with the Axoscope software (both from Axon Instruments, Foster City, CA).

Photoelectric Measurements

Photoelectric responses of *C. reinhardtii* suspensions were recorded in the unilateral mode of a population assay (Sineshchekov et al. 1992, 1994). An array of light-emitting diodes (LEDs) with a 500 nm peak and a 35 nm bandwidth (NSPE510S, Nichia Chemical Industries, Anan, Japan), or a commercial photoflash gun (model 283, Vivitar, Korea) with a broad-band filter with maximum transmittance at 500 nm was used as the light sources. In experiments with the LEDs light pulses of 10 ms duration (1.2×10^{20} photons m^{-2} s^{-1}) were applied in trains of 10 with 1 s intervals between individual pulses and 10 s between successive trains. Data obtained from each individual train were pooled together. In experiments with the photoflash gun, individual flashes were applied with the time interval 60 s between successive flashes. The pCLAMP 6.3 software (Axon Instruments, Foster City, CA) was used for triggering the light stimuli and for data acquisition. Photoelectric responses were fed to a low-noise current amplifier (model 564, Ithaco, Ithaca, NY) in series with and a Digidata 1200 DMA board (Axon Instruments, Foster City, CA). Origin 6.1 software (OriginLab, Northampton, MA) was used for evaluation of the data.

The pH values of the test solutions (tryptone, its fractions and individual amino acids) were adjusted to that of cell suspensions. The test solutions were added as 1/10 of the total volume of a cell sample to ensure rapid equilibration of the test substance concentration. The same volume of pure medium was added to an identical cell sample as control. Each individual experiment consisted of two test and two control measurements performed in a mirror sequence to compensate for a possible change in the culture state over time. The peak amplitude of the net PCs was measured as shown in Figure 3. The values obtained in the control sample were subtracted from those obtained in the test sample to account for a possible effect of buffer addition.

Gel Filtration, Mass Spectrometry and Determination of Amino Nitrogen

Tryptone (30 mg) was dissolved in 2 mL 0.1 N acetic acid, 5% acetonitrile, and loaded on a Bio-Gel P2 (Bio-Rad Laboratories, Hercules, CA) column (1.5 x 95 cm), equilibrated in the same buffer. Peptides were eluted at a flow rate of 0.1 mL/min. Fractions of 12 mL volume were collected, dried and redissolved in 0.5 mL NMM for testing the biological response. Peptides in each fraction were identified by reversed phase high performance liquid chromatography/electrospray ionization tandem mass spectrometry (M.A. Kutuzov, N. Deighton and H.V. Davies, unpublished work). Amino nitrogen was determined with a modified ninhydrin method (Doi et al., 1981).

Results and Discussion

Chemotaxis to NH_4^+ is induced by nitrogen depletion in vegetative cells, whereas chemotaxis to tryptone requires formation of mature gametes

Vegetative *C. reinhardtii* cells in growth medium are not chemotactic. It has been shown earlier that transfer of the cells to nitrogen-free medium induces chemotaxis to both NH_4^+ (Byrne et al., 1992) and tryptone (Govorunova and Sineshchekov, 2003). Nitrogen depletion is also a key signal for gametogenesis in *C. reihardtii* (Sager and Granick, 1954), although formation of mature mating-competent gametes in addition requires light (Treier et al., 1989; Pan et al., 1997; Huang and Beck, 2003). The development of the chemotactic sensitivity to tryptone after the removal of nitrogen was strictly light-dependent (Figure 1A, squares; Govorunova and Sineshchekov, 2003). Its time course closely matched that of formation of mature gametes, which under our conditions was completed after ~14-17 h (Figure 1A, open circles). In contrast, the chemotactic sensitivity to NH_4^+ developed with the same rate in the light and in the dark, reached its maximum ~2 h after the removal of nitrogen, and rapidly returned to zero after that (Figure 1B, squares). This result shows that the development of the chemotactic sensitivity to NH_4^+ is independent of gametogenesis. It therefore appears that the two strongest chemoattractants, tryptone and NH_4^+, are specific for different stages of *C. reinhardtii* life cycle.

Chemotaxis to NH_4^+ of the C3-sil$_6$ strain, in which the content of the C3 subunit of the RNA-binding protein CHLAMY1 was reduced to <25% of the wild type (Iliev et al., 2006), and control strain transformed with an empty vector, was measured under LD conditions (Figure 2, open circle and filled squares, respectively). It has been shown earlier that changes in chemotaxis observed under LD conditions persist in continuous light, and thus reflect the true circadian rhythm (Byrne et al., 1992). In contrast to the rhythms of phototaxis and nitrite reductase activity (Iliev et al., 2006) that were measured under circadian conditions, the acrophase of the chemotaxis rhythm was not shifted upon C3 silencing in a LD regime (data not shown). The magnitude of the response in the C3-sil$_6$ strain was equal to that in the control strain during the night, but was higher during the day (Figure 2). This result suggests that CHLAMY1 may be involved in regulation of the diurnal rhythm of chemotaxis. Future experiments will have to clarify if the shift in acrophase that was observed with the circadian rhythms of phototaxis and nitrite reductase activity in C3-sil strains is also present in the case of the chemtotaxis rhythm under circadian conditions. In case of the mouse *clock* mutant and the *per1* mutant in *C. reinhardtii*, differences in period were only observed under circadian free-running conditions, but not in a LD regime (Antoch et al., 1997; Voytsekh et al., 2008).

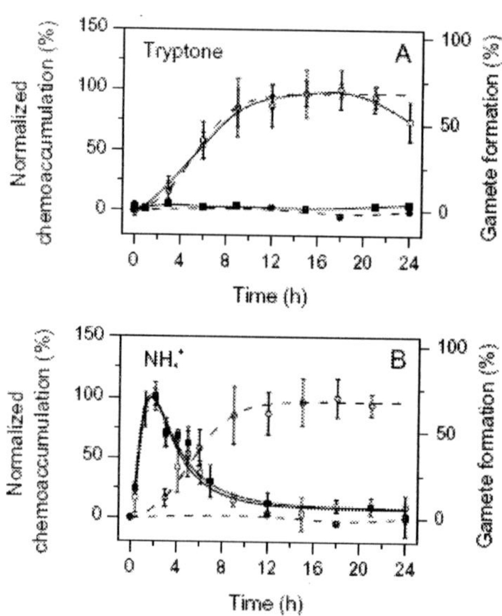

Figure 1. Chemotaxis to NH_4^+ is induced by nitrogen depletion, whereas chemotaxis to tryptone requires formation of mature gametes. Vegetative *C. reinhardtii* cells were transferred to nitrogen-free medium at zero time and incubated in the light (open symbols) or dark (filled symbols). Chemotaxis (squares, left axis) was assayed by measuring chemoaccumulation of the cells by a capillary assay with 1% tryptone (A) or 30 mM NH_4Cl (B). Formation of mature gametes was assayed by measurement of their fusion with mature gametes of the opposite mating type. The data points are the means ± S.E. of 3-6 independent experiments. The curve for chemoaccumulation in response to tryptone is modified from Govorunova and Sineshchekov, 2003, and the curve for formation of mature gametes is from Govorunova et al., 2007

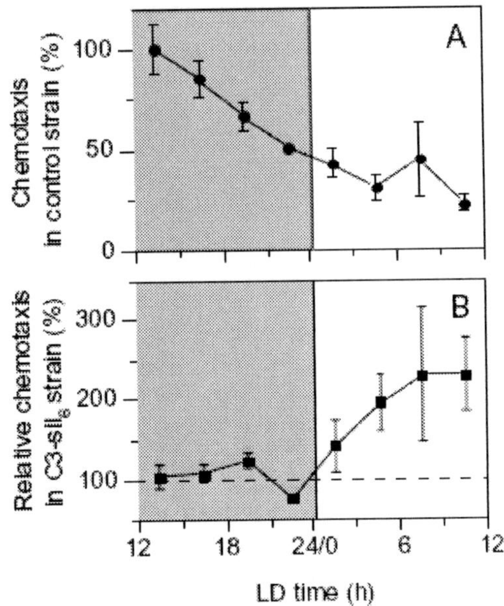

Figure 2. The influence of C3 subunit of RNA-binding protein CHLAMY1 on chemotaxis to NH_4^+ under LD conditions. Chemotaxis was assayed by measuring chemoaccumulation by a capillary assay (30 mM NH_4Cl) in the C3-sil_6 strain with a reduced content of the C3 subunit, and the in control strain transformed with an empty vector. The cells were preincubated in nitrogen-free medium for 1 h before measurements. (A) Chemotaxis in the control strain transformed with an empty vector, normalized to its maximal value obtained during the period of measurement; (B) relative chemotaxis in the C3-sil_6 strain with a reduced content of the C3 subunit, normalized to the value measured in the control strain at the same time point. Grey background shows the dark part of the 12 h day/12 h night cycle. Zero time is the beginning of the day phase during the period of the measurements.

Inhibition of PCs by Tryptone Depends on Extracellular K^+

When suspensions of *C. reinhardtii* gametes are excited with brief flashes of light, transient PCs can be recorded with appropriately placed electrodes (Figure 3). These currents are strongly inhibited by the addition of the chemoeffector tryptone, and the degree of this inhibition correlates with the strength of chemotactic accumulation of the gametes upon variation of experimental conditions (Govorunova and Sineshchekov, 2003).

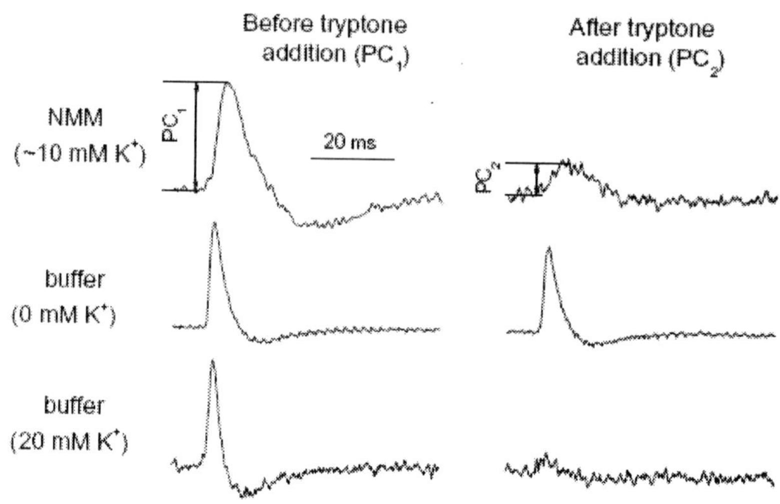

Figure 3. Tryptone-induced inhibition of photoelectric currents in suspensions of *C. reinhardtii* gametes depends on extracellular K^+. The traces were recorded before and 1 min after the addition of tryptone (0.1% final concentration) to gametes preincubated in NMM or Tris-HCl buffer at the indicated external K^+ concentration. PC_1 is the peak value of the net photoreceptor currents recorded before the addition of tryptone. PC_2 is the peak value of the net photoreceptor currents recorded after the addition of tryptone. The current amplitudes were normalized with respect to PC_1

The addition of tryptone (0.1% final concentration) to gametes in standard NMM medium (~10 mM K^+, for the exact composition see Materials and Methods) resulted in a ~3-fold decrease in the amplitude of the net flash-induced PCs (Figure 3, top). One possible explanation for this observation would be that the chemoeffector tryptone induces depolarizing current across the cell membrane, which inhibits PCs. In order to verify this hypothesis, we studied the dependence of the tryptone-induced inhibition of PCs on the ionic composition of the external medium. When gametes were transferred from NMM to measuring buffer containing no monovalent cations besides Tris and H^+ (100 μM $CaCl_2$, 5 mM Tris-HCl, pH 8.5), the addition of the same tryptone concentration decreased the amplitude of the currents only to 80% of the original level (Figure 3, middle). If buffer was supplemented with KCl, the response to tryptone was restored (Figure 3, bottom).

The time course of the response to tryptone upon repetitive flash photoexcitation showed a rapid initial drop in the amplitude of the net PCs, followed by a slow recovery. An increase in the extracellular K^+ concentration both increased the magnitude of the initial drop and slowed down the recovery

(Figure 4). The time resolution of our technique is limited by the time of the dark recovery of PCs (~1 min after a saturating flash, Sineshchekov et al., 1990; Govorunova et al., 1997), which is too low for reliable measurement of the magnitude of the initial drop. Therefore, for the purpose of this study we did not attempt to separate the influence of tested experimental conditions on the initial drop and subsequent recovery of the current amplitude, but regarded the effect of tryptone as integral. According to published estimations (Malhotra and Glass, 1995), $\Delta\mu K^+$ in *C. reinhardtii* is negative at extracellular K^+ concentrations above 0.3 mM. Therefore, we hypothesize that the observed dependence of the response to tryptone on extracellular K^+ signifies that tryptone induces a K^+ influx, which is enhanced at high extracellular K^+. However, we cannot rule out that, in fact, tryptone induces an influx of a different cation species (such as Ca^{2+}), or efflux of anions, followed by a K^+ efflux during the recovery phase of the response. This efflux would be suppressed at high extracellular K^+, which would be an alternative explanation of the increased integral response to tryptone under these conditions. The development of more direct methods for probing chemosensory transduction with an improved time resolution is needed to discriminate between these two possibilities.

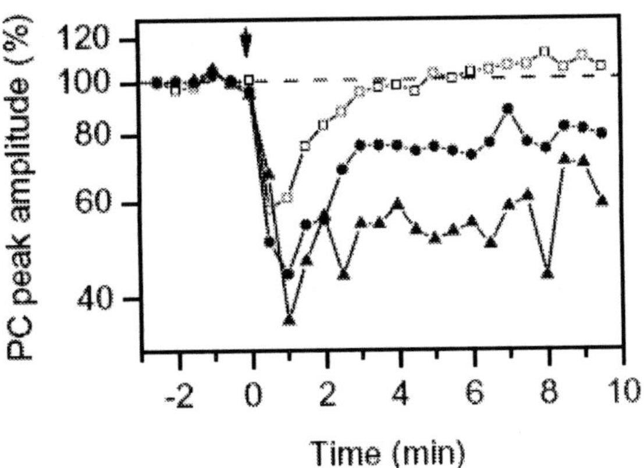

Figure 4. The time course of the responses to tryptone at different extracellular K^+ concentrations. The peak amplitude of the net PCs was measured upon repetitive flash excitation at 0 (open squares), 10 (filled circles) and 30 (filled triangles) mM extracellular K^+. Tryptone (final concentration 0.1%) was added at time zero, as indicated by the arrow. The data points are individual values from a typical experiment

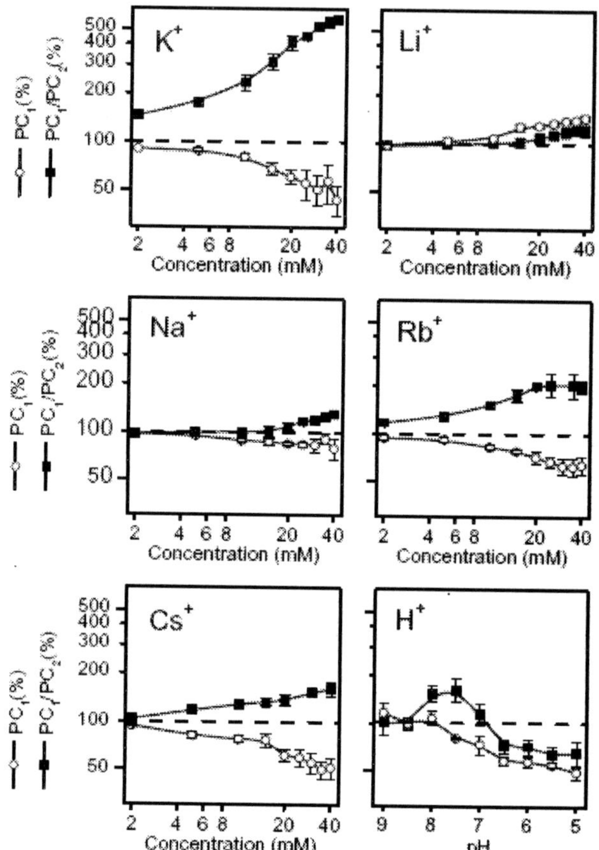

Figure 5. The influence of monovalent metal cations and pH on the net PCs and the response to tryptone. The peak amplitude of the net PCs before the addition of tryptone (PC_1, open circles) and the response to tryptone (PC_1/PC_2; filled squares) were measured as shown in Figure 3. The values were normalized to those measured in cation-free buffer (pH 8.5). The dependence on the pH was measured at 10 mM extracellular K^+. The final concentration of tryptone was 0.1%. The data points are the means ± S.E. of 4 to 5 independent experiments

To quantify the integral response to tryptone, we averaged the amplitudes of PCs measured upon three subsequent flashes before (PC_1), and three flashes after the addition of tryptone (PC_2), and calculated their ratio, PC_1/PC_2. This ratio shows how much the amplitude of the net PCs is inhibited by tryptone, i.e. the higher this ratio, the stronger the response to tryptone. It gradually increased with an increase in the external K^+ concentration up to 40 mM, above which the signal-to-noise ratio became too low for reliable measurement of the current amplitude (Figure 5, filled squares). An increase in the concentration of other monovalent metal cations also increased the response to tryptone, although less efficiently than K^+ in the order Rb^+>Cs^+>Na^+>Li^+ (Figure 5, filled squares). It is noteworthy that in the absence of tryptone the addition of all tested metal cations but Li^+ decreased the amplitude of PCs (Figure 5, open circles). These results show that a chemosensory system involves a specific set of ion channels, different from those activated by photoexcitation of channelrhodopsins.

The dependence of the response to tryptone on the external pH (Figure 5, filled squares) is consistent with our hypothesis that chemosensory transduction in *C. reinhardtii* involves depolarizing current. The negative resting potential in *C. reinhardtii* linearly increases upon an increase in the external pH from 3 to 9 (Malhotra and Glass, 1995). Depolarizing current would be facilitated by hyperpolarization of the cell membrane. Indeed, the response to tryptone increased upon an increase in the pH from 5 to 7.5. Its decline above this value may be due to reasons other than changes in the membrane resting potential. For instance, possible changes in the protonation state of the active substance(s) in tryptone at the high pH may interfere with their binding to receptors.

The selectivity sequence K^+>Rb^+>Cs^+>Na^+>Li^+, known as the Eisenman sequence IV (Eisenman and Horn, 1983), is rather unusual for plant K^+ channels. Most of them are blocked by Cs^+, rather than are permeable for it (Hedrich and Dietrich, 1996). We tested pharmacological properties of putative K^+ channels that mediate chemosensory transduction in *C. reinhardtii* by measuring the response to tryptone at 10 mM extracellular K^+. As expected, Cs^+ up to 10 mM did not block the response (data not shown). Also inefficient were 10 μM Al^{3+}, a blocker of uptake K^+ channels in guard cells (Schroeder, 1988), and 1 mM 4-aminopyridine, an inhibitor of delayed rectifier K^+ channels in animal cells (Nelson and Quayle, 1995). In contrast, extracellular Ba^{2+} and Ca^{2+} in millimolar concentrations blocked the response to tryptone (Figure 6). Such blockade is characteristic of many K^+ channels in charophytes and higher plants (Kitasato, 1973; Schroeder et al., 1987; Fairley-Grenot and Assmann, 1992; Wegner et al., 1994; Fan et al., 2001). In *C. reinhardtii*, K^+ channels were implicated in repolarization of the membrane after depolarization by PCs (Nonnengässer et al., 1996; Govorunova et al., 1997). These channels were also blocked by extracellular Ca^{2+}, but it is not clear whether the same channel species is involved in the response to tryptone.

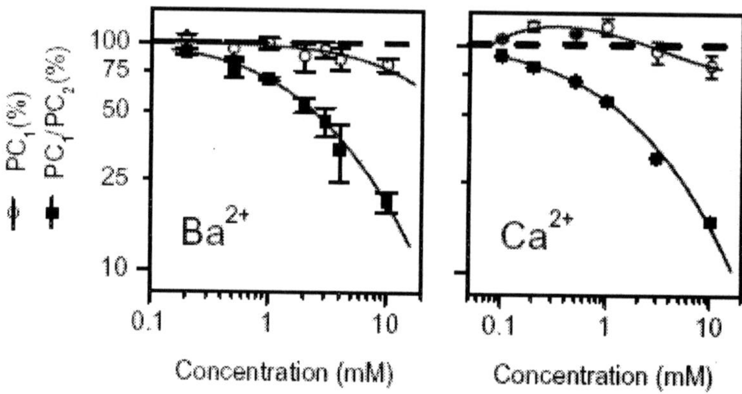

Figure 6. Blockade of the response to tryptone by Ba^{2+} and Ca^{2+}. The peak amplitude of the net PCs before the addition of tryptone (PC_1, open circles) and the response to tryptone (PC_1/PC_2; filled squares) were measured as shown in Figure 3. The values were normalized to those measured in the absence of the blocking cations. The final concentration of tryptone was 0.1%. The data points are the means ± S.E. of 4 to 6 independent experiments

We have shown earlier that the response to tryptone was suppressed by 3-isobutyl-1-methylxanthine (IBMX), an inhibitor of cyclic nucleotide phosphodiesterases (PDE) (Govorunova et al., 2007). This result suggests that cyclic nucleotides may act as second messengers in chemosensory transduction in *C. reinhardtii* by modulation of K^+ channels activity. But, xanthines are also known to directly inhibit K^+ channels and induce a release of Ca^{2+} from intracellular stores (Usachev et al., 1995, Zhao et al., 2002). Therefore, we tested whether preincubation of gametes with dibutyryl-cAMP (db-cAMP) changed their response to tryptone. In db-cAMP-preincubated gametes, the addition of tryptone elicited a weaker response, as compared to control (Govorunova et al., 2007; Figure 7). This result corroborated our hypothesis that cAMP may mediate chemotactic response to tryptone in *C. reinhardtii*. Direct binding of regulatory cyclic nucleotides to K^+ channels or an indirect control, for instance, via activation of cyclic nucleotide-dependent protein kinases, can be suggested as possible biochemical mechanisms of *C. reinhardtii* chemosensing. *C. reinhardtii* genome mining (http://genome.jgi-psf.org/chlamy) returns several candidates for both cyclic-nucleotide gated K^+ channel and cAMP-specific PDE genes, but further studies are needed to test whether they regulate *C. reinhardtii* chemotaxis.

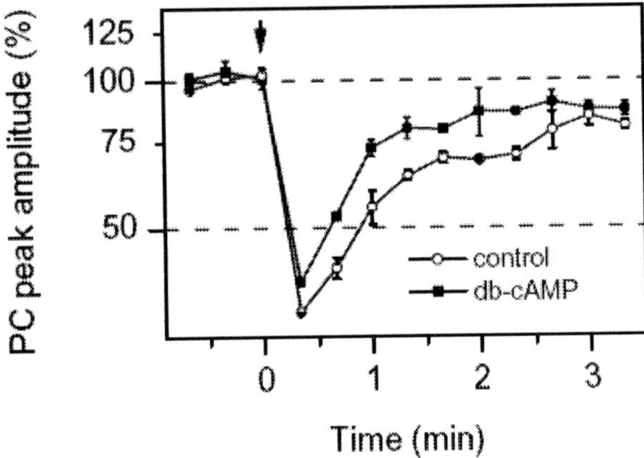

Figure 7. The influence of db-cAMP on the response to tryptone. The peak amplitude of the net PCs was measured upon repetitive flash excitation in gametes preincubated for 1 h with 10 mM final concentration of NaCl (open circles) or db-cAMP sodium salt (filled squares). Tryptone (final concentration 0.03%) was added at time zero, as indicated by the arrow. The data are the means ± S.E. of 4 independent experiments

Analysis of Tryptone Fractions

As the first step toward identification of an active substance(s) responsible for the chemoattractant action of tryptone we examined the functional efficiency of tryptone fractions separated by gel filtration. The fractions were tested in cells obtained by preincubation in NMM upon illumination (mature gametes) or in the dark (nitrogen-deprived vegetative cells), as only the former responded to whole tryptone. Only fractions 10 and 11 inhibited the currents in mature gametes (i.e., PC_1/PC_2 ratio was > 100%), whereas all other fractions stimulated the currents in both cell preparations, as did tryptone in nitrogen-deprived vegetative cells (PC_1/PC_2 ratio was < 100%) (Figure 8, top). Each tryptone fraction was subjected to mass spectrometry analysis for oligopeptide identification. However, our system could only identify oligopeptides composed of three or more amino acid residues, and no such substances were found in the two active fractions (Figure 8, middle). Therefore, it can be concluded that active substances must be molecules not larger than tripeptides. These are most likely dipeptides and/or individual amino acids, as shows the presence of amino nitrogen in the active fractions (Figure 8, bottom). The two active fractions contained amino nitrogen in concentrations 54 and 38 μg mL$^-$

^{1}, respectively. For a molecule with M_r = 100, this corresponds to the final concentrations of 54 and 38 μM, respectively, which gives us an upper limit for the chemotactic sensitivity of *C. reinhardtii*.

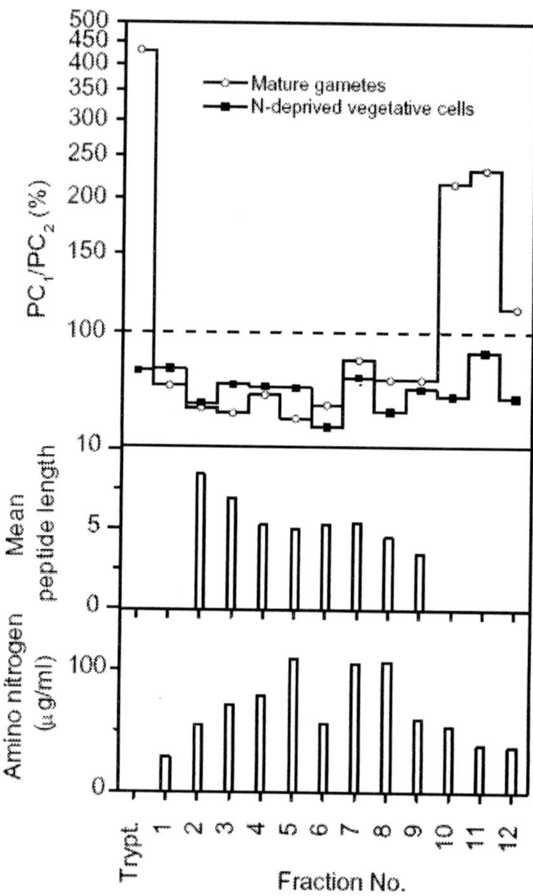

Figure 8. Analysis of tryptone fractions. Top, responses to individual tryptone fractions obtained by gel filtration, defined as PC_1/PC_2 ratio. PC1 and PC2 were measured as shown in Figure 3. The response to tryptone (0.1% final concentration) is shown for comparison. Mature gametes (open circles) and nitrogen-deprived vegetative cells (filled squares) were produced by overnight incubation in NMM in the light or in the dark, respectively. Middle, the mean lengths of peptides identified in the tryptone fractions by mass spectrometry. Bottom, amino nitrogen content of the fractions. The data are the means of independent measurements in two sets of fractions obtained by separate gel filtration procedures on different days

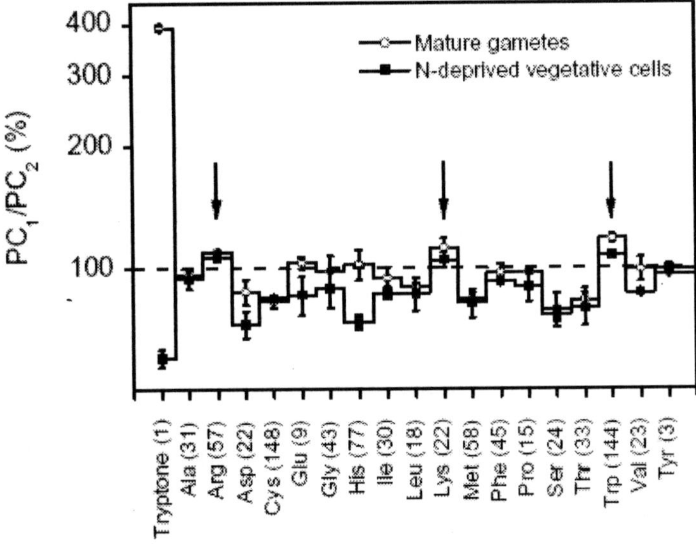

Figure 9. Responses to individual amino acids. Top, the response to individual tryptone fractions obtained by gel filtration, defined as PC_1/PC_2 ratio. PC_1 and PC_2 were measured as shown in Figure 3. Mature gametes (open circles) and nitrogen-deprived vegetative cells (filled squares) were produced by overnight incubation in NMM in the light or in the dark, respectively. The final concentrations of Trp and Tyr were 0.2 and 5 mM, respectively, those of all other amino acids – 10 mM. These concentrations exceeded by the factor written in parentheses after the name of each amino acid the concentrations of the respective amino acids in 0.1% tryptone, the response to which is shown for comparison. Data on amino acid composition of tryptone are from the Difco manual (http://www.bd.com/industrial/difco/manual.asp). The data are the means ± S.E. of 4 independent experiments

We also tested responses to 18 individual common amino acids (Figure 9). None of them elicited a significant inhibitory response in gametes and stimulatory in nitrogen-deprived vegetative cells, as it was typical of tryptone or its active fractions. Arg, Lys and Trp slightly inhibited PCs in mature gametes ($PC_1/PC_2 > 100\%$), but, in contrast to tryptone, they also did so in nitrogen-deprived vegetative cells. Most other amino acids stimulated PC ($PC_1/PC_2 < 100\%$) in both tested types of culture, as tryptone did in nitrogen-deprived vegetative cells. Interestingly, Arg is the only amino acid reported to induce chemotaxis in vegetative cells of *C. reinhardtii* (Hirschberg and Rogers, 1978). Arg is also the only amino acid that is taken up by *C. reinhardtii* in the absence of acetate, i.e., under conditions of our experiments

(Kirk and Kirk, 1978). It is however unlikely that this transport system is involved in chemotaxis, because Lys and Trp, which are not transported, also inhibit PCs in gametes. Two hypotheses can be put forward to explain why individual amino acids do not mimic chemoattractant properties of tryptone. First, chemotaxis in *C. reinhardtii* may require a synergistic action of individual amino acids, as it has been found in zoospores of the water mold *Allomyces* (Machlis, 1969). Second, a yet unidentified substance(s), possibly a dipeptide(s), can be responsible for the chemoattractant action of tryptone on *C. reinhardtii* gametes. Further research is needed to test these hypotheses.

Conclusion

The sensitivities to the two strongest known chemoattractants, protein hydrolysates and NH_4^+ ions, develop at different stages of the *C. reinhardtii* life cycle and involve different signaling mechanisms. Chemotaxis to NH_4^+ transiently develops in vegetative cells upon nitrogen deprivation. The amplitude of its diurnal rhythm is likely to be regulated by the clock-controlled RNA-binding protein CHLAMY1. The gamete-specific chemotaxis to casein hydrolysate (tryptone) is likely mediated by cAMP-regulated K^+ currents across the plasma membrane. Active substances in tryptone responsible for its chemoattractant action on *C. reinhardtii* gametes are compounds of low molecular weight, possibly, dipeptides and/or a mixture of individual amino acids.

Acknowledgments

This work was supported by the Russian Foundation for Basic Research grant 08-04-01453 and an Institutional Partnership grant from the Alexander von Humboldt Foundation RUS/1009059. We thank Heather Ross (Scottish Crop Research Institute, Dundee, Scotland, UK) for the use of FPLC.

References

Antoch MP; Song. EJ; Chang AM; Vitaterna MH; Zhao Y; Wilsbacher LD; Sangoram AM; King DP; Pinto LH; Takahashi JS. (1997) Functional identification of the mouse circadian clock gene by transgenic BAC rescue. *Cell*, 89, 655-667

Beck CF, Haring MA (1996) Gametic differentiation of *Chlamydomonas. Int Rev Cytol*, 168, 259-302

Byrne TE, Wells MR, Johnson CH (1992) Circadian rhythms of chemotaxis to ammonium and of methylammonium uptake in *Chlamydomonas. Plant Physiol*, 98, 879-886

Doi E, Shibata D, Matoba T (1981) Modified colorimetric ninhydrin methods for peptidase assay. *Anal Biochem*, 118, 173-184

Eisenman G, Horn R (1983) Ionic selectivity revisited: the role of kinetic and equilibrium processes in ion permeation through channels. *J Membr Biol*, 76, 197-225

Ermilova EV, Zalutskaya ZM, Lapina TV, Nikitin MM (2003) Chemotactic behavior of *Chlamydomonas reinhardtii* is altered during gametogenesis. *Curr Microbiol,* 46, 261-226

Fairley-Grenot KA, Assmann SM (1992) Permeation of Ca^{2+} through K^{+} channels in the plasma membrane of *Vicia faba* guard cells. *J Membr Biol*, 128, 103-113

Fan LM, Wang YF, Wang H, Wu WH (2001) In vitro *Arabidopsis* pollen germination and characterization of the inward potassium currents in *Arabidopsis* pollen grain protoplasts. *J Exp Bot*, 52, 1603-1614

Govorunova EG, Sineshchekov OA, Hegemann P (1997) Desensitization and dark recovery of the photoreceptor current in *Chlamydomonas reinhardtii. Plant Physiol*, 115, 633-642

Govorunova EG, Sineshchekov OA (2003) *Integration of photo- and chemosensory signaling pathways in Chlamydomonas Planta*, 216, 535-540

Govorunova EG, Jung K-W, Sineshchekov OA, Spudich JL (2004) *Chlamydomonas* sensory rhodopsins A and B: Cellular content and role in photophobic responses *Biophys J*, 86, 2342-2349

Govorunova EG, Sineshchekov OA (2005) *Chemotaxis in the green flagellate alga Chlamydomonas Biochemistry* (Mosc), 70, 717-725

Govorunova EG, Voytsekh OO, Sineshchekov OA (2007) Changes in photoreceptor currents and their sensitivity to the chemoeffector tryptone during gamete mating in *Chlamydomonas reinhardtii Planta*, 225, 441-449

Harris EH (1989) The *Chlamydomonas* source book: A comprehensive guide to biology and laboratory use. Academic Press, San Diego

Harz H, Hegemann P (1991) Rhodopsin-regulated calcium currents in *Chlamydomonas*. *Nature*, 351, 489-491

Hedrich R, Dietrich P (1996) Plant K^+ channels: similarity and diversity. *Bot Acta* 109, 94-101

Hegemann P, Berthold P (2009) Sensory photoreceptors and light control of flagellar activity In The *Chlamydomonas* Sourcebook, 2nd Edition, v. 3 *Cell Motility and Behavior,* GB Witman, ed, Elsevier, Amsterdam, 395-429

Hirschberg R, Rodgers S (1978) Chemoresponses of *Chlamydomonas reinhardtii*. *J Bacteriol*, 134, 671-673

Huang K, Beck CF (2003) Phototropin is the blue-light receptor that controls multiple steps in the sexual life cycle of the green alga *Chlamydomonas reinhardtii*. *Proc Natl Acad Sci* USA, 100, 6269-6274

Iliev D, Voytsekh O, Schmidt EM, Fiedler M, Nykytenko A, Mittag M (2006) A heteromeric RNA-binding protein is involved in maintaining acrophase and period of the circadian clock. *Plant Physiol*, 142, 797-806

Johnson CH (2001) Endogenous timekeepers in photosynthetic organisms. *Annu Rev Physiol*, 63, 695-728

Kitasato H (1973) K^+ permeability of *Nitella clavata* in the depolarized state. *J Gen Physiol*, 62, 535-549

Kirk DL, Kirk MM (1978) Carrier mediated uptake of arginine and urea by *Chlamydomonas reinhardtii*. *Plant Physiol*, 61, 556-560

Kreimer G (2009) The green algal eyespot apparatus: a primordial visual system and more? *Curr Genet*, 55, 19-43

Litvin FF, Sineshchekov OA, Sineshchekov VA (1978) Photoreceptor electric potential in the phototaxis of the alga *Haematococcus pluvialis*. *Nature,* 271, 476-478

Machlis L (1969) Zoospore chemotaxis in the watermold *Allomyces*. *Physiol Plantarum,* 22, 126-139

Malhotra B, Glass ADM (1995) Potassium fluxes in *Chlamydomonas reinhardtii*. 1. Kinetics and electrical potentials. *Plant Physiol*, 108, 1527-1536

Matsuo T, Okamoto K, Onai K, Niwa Y, Shimogawara K, Ishiura M (2008) A systematic forward genetic analysis identified components of the *Chlamydomonas* circadian system. *Genes Dev,* 22, 918-930

Mittag M (1996) Conserved circadian elements in phylogenetically diverse algae. *Proc Natl Acad Sci USA,* 93, 14401-14404

Mittag M and Wagner V (2003) The circadian clock of the unicellular eukaryotic model organism *Chlamydomonas reinhardtii. Biol Chem*, 384, 689-695

Mittag M, Kiaulehn S, Johnson CH (2005) The circadian clock in *Chlamydomonas reinhardtii*. What is it for? What is it similar to? *Plant Physiol,* 137, 399-409

Nagel G, Ollig D, Fuhrmann M, Kateriya S, Musti, Bamberg E, Hegemann P (2002) Channelrhodopsin-1: a light-gated proton channel in green algae, *Science*, 296, 2395-2398

Nagel G, Szellas T, Huhn W, Kateriya S, Adeishvili N, Berthold P, Ollig D, Hegemann P, Bamberg E (2003) Channelrhodopsin-2, a directly light-gated cation-selective membrane channel. *Proc Natl Acad Sci* USA, 100, 13940-13945

Nelson MT, Quayle JM (1995) Physiological roles and properties of potassium channels in arterial smooth muscle. *Am J Physiol,* 268, C799-C822

Nonnengässer C, Holland E-M, Harz H, Hegemann P (1996) The nature of rhodopsin-activated photocurrents in *Chlamydomonas*. II. *Influence of monovalent ions Biophys J*, 70, 932-938

Pan J-M, Haring MA, Beck C-F (1997) Characterization of blue light signal transduction chains that control development and maintenance of sexual competence in *Chlamydomonas reinhardtii Plant Physiol*, 115, 1241-1249

Sager R, Granick S (1954) Nutritional control of sexuality in *Chlamydomonas reinhardtii*. J Gen Physiol, 37, 729-742

Schmidt M, Gessner G, Luff M, Heiland I, Wagner V, Kaminski M, Geimer S, Eitzinger N, Reissenweber T, Voytsekh O, Fiedler M, Mittag M, Kreimer G (2006) Proteomic analysis of the eyespot of *Chlamydomonas reinhardtii* provides novel insights into its components and tactic movements. *Plant Cell,* 18, 1908-1930

Schroeder JI (1988) K^+ -transport properties of K^+-channels in the plasma membrane of *Vicia faba* guard cells. *J Gen Physiol*, 92, 667-683

Schroeder JI, Raschke K, Neher E (1987) Voltage dependence of K^+ channels in guard-cell protoplasts. *Proc Natl Acad Sci USA,* 84, 4108-4112

Schulze T, Prager K, Dathe H, Kelm J, Kießling P, Mittag M (2010) How the green alga *Chlamydomonas reinhardtii* keeps time. *Protoplasma*, in press

Serrano G, Herrera-Palau R, Romero JM, Serrano A, Coupland G, Valverde F (2009) *Chlamydomonas* CONSTANS and the evolution of plant photoperiodic signaling. *Curr Biol*, 19, 359-368

Sineshchekov OA, Litvin FF, Keszthelyi L (1990) Two components of photoreceptor potential of the flagellated green alga *Haematococcus pluvialis*. *Biophys* J, 57, 33-39

Sineshchekov OA, Govorunova EG, Der A, Keszthelyi L, Nultsch W (1992) Photoelectric responses in phototactic flagellated algae measured in cell suspension. *J Photochem Photobiol B: Biol*, 13, 119-134

Sineshchekov OA, Govorunova EG, Der A, Keszthelyi L, Nultsch W (1994) Photoinduced electric currents in carotenoid-deficient *Chlamydomonas* mutants reconstituted with retinal and its analogs. *Biophys J,* 66, 2073-2084

Sineshchekov OA, Jung K-H, Spudich JL (2002) Two rhodopsins mediate phototaxis to low- and high-intensity light in *Chlamydomonas reinhardtii*. *Proc Natl Acad Sci* USA, 99, 8689-8694

Sineshchekov OA, Spudich JL (2005) Sensory rhodopsin signaling in green flagellate algae In *Handbook of Photosensory Receptors*, Briggs W.R., Spudich J.L., eds, Wiley-VCH, Weinheim, 25-42

Sjoblad RD, Frederikse PH (1981) Chemotactic responses of *Chlamydomonas reinhardtii. Mol Cell Biol*, 1, 1057-1060

Treier U, Fuchs S, Weber M, Wakarchuk WW, Beck CF (1989) Gametic differentiation in *Chlamydomonas reinhardtii*: light dependence and gene expression patterns. *Arch Microbiol*, 152, 572-577

Usachev Y, Kostyuk P, Verkhratsky A (1995) 3-Isobutyl-1-methylxanthine (IBMX) affects potassium permeability in rat sensory neurones via pathways that are sensitive and insensitive to $[Ca^{2+}]_{in}$. *Pflugers Arch.* 430, 420-428

Voytsekh O, Seitz SB, Iliev D, Mittag, M (2008) Both subunits of the circadian RNA-binding protein CHLAMY1 can integrate temperature information. *Plant Physiol*, 147, 2179-2193

Wegner LH, De Boer AH, Raschke, K. (1994) Properties of the K^+ inward rectifier in the plasma membrane of xylem parenchyma cells from barley roots: effects of TEA^+, Ca^{2+}, Ba^{2+} and La^{3+}. *J Membr Biol*, 142, 363-379

Zhao FL, Lu SG, Herness S (2002) Dual actions of caffeine on voltage-dependent currents and intracellular calcium in taste receptor cells. *Am J Physiol Regul Integr Comp Physiol*, 283, R115-R129

Zhao B, Schneid C, Iliev D, Schmidt EM, Wagner V, Wollnik F, Mittag M (2004) The circadian RNA-binding protein CHLAMY 1 represents a novel type heteromer of RNA recognition motif and lysine homology domain-containing subunits. *Eukaryot Cell*, 3, 815-825

In: Chemotaxis: Types, Clinical Significance... ISBN: 978-1-61728-495-3
Editors: T. C. Williams, pp. 157-187

Chapter 6

Chemotaxis-Based Assay for the Biological Action of Silver Nanoparticles

E. M.Egorova*[*1], *S. I. Beylina*[≠2], *N. B.Matveeva*[2] and *L. S. Sosenkova*[1]

[1] Institute of General Pathology and Pathophysiology, Russian Academy of Medical Sciences, Moscow, Russia

[2] Institute of Theoretical and Experimental Biophysics, Russian Academy of Sciences, Pushchino, Moscow region, Russia

Abstract

We propose a new way for testing the biological action of silver nanoparticles. The nanoparticles (with sizes ranging from 3 to 16 nm) were obtained by the original method of biochemical synthesis in reverse micelles stabilized by anionic surfactant bis-(2-ethylhexyl) sodium sulfosuccinate (AOT). From micellar solution the nanoparticles were transferred into the water phase; water solutions of the nanoparticles were used for testing their biological activity. Our assay is based on negative chemotaxis, a motile reaction of cells to an unfavorable chemical

* emenano@mail.ru

≠ beylina@iteb.ru

environment. Plasmodium of the slime mold *Physarum polycephalum* used as an object is a multinuclear amoeboid cell with unlimited growth and auto-oscillatory mode of locomotion. Biocidal and repellent effects were compared for silver nanoparticles, Ag^+ ions, and AOT; the latter two reagents were introduced in the concentrations equal to those present in the nanoparticles' solution. All substances were tested in water solution and in the agar gel. We have revealed that in characteristics common for repellents, such as increase of the period of contractile auto-oscillations, decrease of the area of spreading on substrate, and substrate preference in spatial tests, silver nanoparticles proved to be substantially more effective than Ag^+ ions, AOT and the sum Ag^+ + AOT. The lethal concentration of the nanoparticles for macroplasmodium in water solution was found to be about 10 μg/ml, the concentrations effective for chemotaxis were 30 times lower. The chemotactic tests allow the quantitative estimation of the biological reaction and monitoring of its dynamics; in resolution, they are superior to the tests based on the lethal action of biocidal agents. It is shown also that the spatial chemotactic tests are sensitive enough to distinguish between the effectiveness of different nanoparticle preparations. In particular, we found that, at the equal silver concentration, the repellent activity is higher for the 5 nm than for the 9 nm particles.

The results obtained allow to conclude that the chemotaxis-based assay could be helpful for finding a proper balance between the efficiency of silver nanoparticles as antimicrobial drug and the risk of tissue damage during medical treatment, and, hence, for elaborating an optimal protocol of their clinical application.

Introduction

In the last years considerable attention is paid to the studies of the biological effects of metal, most often silver nanoparticles. The main reason is that these nanoparticles have found already a number of practical applications, connected mainly with their antimicrobial properties. Silver nanoparticles are used both as water solutions (as drugs, e.g. colloidal silver) and for the modification of various materials which acquire the biocidal activity. Here belong various health-care products for medical purposes (wound dressings, antimicrobial bandages et al) [1] and for the everyday use (cosmetics, clothes, washing machines, toys et al). The authorities responsible for public health now face the problem with control of the safety of nanoparticles-based products from various manufacturers (e.g.[2]) since, on the one hand, the evidence exists on the human pathologies provoked by the penetration of nanoparticles and, on the other hand, no reliable safety standards can be elaborated today because of the insufficient knowledge on the extent of toxicity and on the mechanism of the nanoparticles interaction with living organisms [2-5]. Therefore, the necessity arose to determine, first, the range of nanoparticle concentrations where their toxic action is dangerous for life and health and, second, to obtain as much information as possible about the effect of various nanoparticle parameters (size, form, structure, surface charge) on their biological activity.

In studies of the biological effects of silver nanoparticles experimental evidence was gained on their toxicity towards bacteria and viruses [6-10], mammalian cultured cells [11,12], fish embryos [13] and animal organisms [14,15]. Apart from the determination of lethal concentrations and threshold concentrations, which mark the appearance of the toxic effects, it was found also that the extent of toxicity depends of the particle size [8,9] and shape [16]. As for the mechanism of the nanoparticles action, some authors believe that it is mediated by the Ag^{+} ions released from the particle surface [8], the other suggest the crucial role of the nanoparticles adsorption on the membrane surface [9,10]. However, version about the role of Ag^{+} ions does not agree with some recent data showing that the effect of nanoparticles is significantly stronger than that of the equivalent concentrations of Ag^{+} ions [11-13].

In our studies of the interaction of silver nanoparticles with biological objects it was confirmed that the nanoparticles possess a high biocidal activity against bacteria and viruses [17] and demonstrate noticeable toxic effects towards more complex systems such as acellular slime mold [18,19], unicellular algae, plant seeds and animals in vivo [17,20,21]. In agreement with the above mentioned data present in literature [11-13] it was found also that, in all cases studied except for the plant seeds, the effect of nanoparticles exceeds that of Ag^{+} ions.

Among the biological objects chosen for our investigations the most detailed analysis of the nanoparticles - living organism interaction was found possible with plasmodium of the acellular slime mold *Physarum polycephalum* because of its capacity for negative chemotactic response, a motile reaction to the unfavorable chemical environment. *Physarum* is a classical object for studying cell motility, and characteristic features of its chemotactic behavior are well established [22–26]. In particular, substances causing negative taxis (repellents) were shown to increase the period of contractile oscillations and to decrease the area of spreading when applied at spatially uniform concentrations [22, 25, 26]. Due to this characteristic features, our experiments with *Physarum* allowed to register not only the lethal events, as with the majority of other objects, but also the changes in its behavior taking place in the mild conditions, at sublethal nanoparticle concentrations. Hence, this chemotaxis – based assay may be regarded as a highly sensitive tool for the studies of the biological action of silver nanoparticles.

In this paper we give a detailed description of the results obtained during the last five years in studies of the interaction of silver nanoparticles with plasmodium of the *Physarum polycephalum.* Our account embraces both the data already published [18, 19] and those found in the more recent experiments. The nanoparticles were obtained by the original method of biochemical synthesis [27] which allows to prepare silver, gold, copper, zinc, etc. particles small in size (below 20 nm), stable in solution for a long time. For example, silver nanoparticles may be preserved in solution for three – five years without noticeable changes of the absorption band position and particle size distribution. In our research five samples of silver nanoparticles in water solution were used; the preparations differed either in the concentration of the solution components or also in the particle size. Both biocidal and repellent effects were studied. In the control experiments, *Physarum* response on the equivalent concentrations of anionic surfactant (present in solution as the nanoparticles' stabilizer) and silver ions was registered. Also the experiments performed with pairs of preparations showed that the chemotactic tests were sensitive enough for recognizing the difference in their efficiency, the fact that allowed to use such tests for the estimation of the particle size effect in the repellent action of silver nanoparticles.

Analysis of the results allowed, as we believe, to derive some valuable information about the toxicity limits, mechanism of the biological action and possibilities of medical applications of silver nanoparticles.

Experimental

Silver Nanoparticles (SNP) were obtained by the biochemical synthesis based on the reduction of silver ions in AOT reverse micelles by the natural flavonoid quercetin (for details see [28, 29]). AOT (sodium bis(2-ethylhexyl)sulfosuccinate) is anionic surfactant widely applied for the formation of reverse micelles. Here we used the AOT purchased from Acros (USA). SNP solutions in distilled water were prepared from their reverse-micellar (further micellar) solutions by the specially developed procedure [30]. Thus obtained SNP water solution (pH 7.6-8.5) contains the nanoparticles in protective shell (AOT bilayer) and some excess of AOT in molecular or aggregated form. In optical absorption spectra, the nanoparticles absorption band lies at 410-415 nm [17, 30].

In our studies five SNP water solutions were tested, containing various SNP and AOT concentrations; the effect of particle sizes was also examined. Parameters of the SNP preparations used are given below.

1. SNP 1- C(Ag) = 1 mg-ion/l, C(AOT) = 5.6 mM
2. SNP2 - C(Ag) = 1.5 mg-ion/l, C(AOT) = 13.5 mM
3. SNP3 - C(Ag) = 1.715 mg-ion/l, C(AOT) = 6.75 mM
4. SNP4 - C(Ag) = 1. 25 mg-ion/l, C(AOT) = 21 mM
5. SNP5 - C(Ag) = 1. 25 mg-ion/l, C(AOT) = 11 mM

The nanoparticles concentration C(SNP) is expressed in the corresponding values of Ag^+ ions concentration, i.e. in mg-ion/l (below also in µg-ion/l), for the convenience of determination of the equivalent Ag^+ concentrations. To express the SNP concentration in mg/ml or µg/ml of silver these values should be multiplied by 108 (the atomic weight of silver); for example, 1 mg-ion/l = 108 mg/l = 0.108mg/ml = 108 µg/ml. We use this scale for the comparison with biological effects of silver nanoparticles reported in literature (see below table 1).

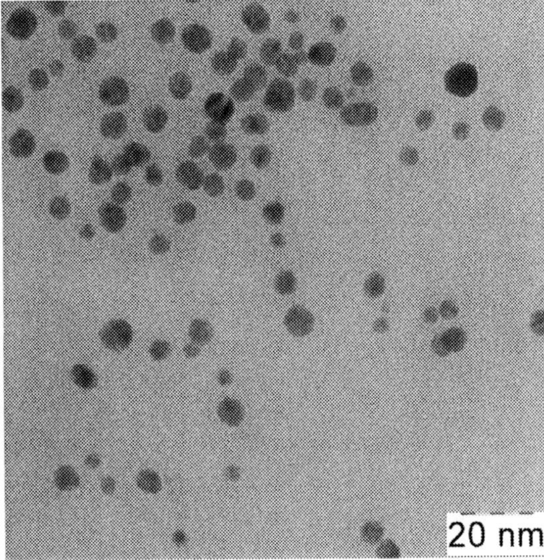

Figure 1 (Continued)

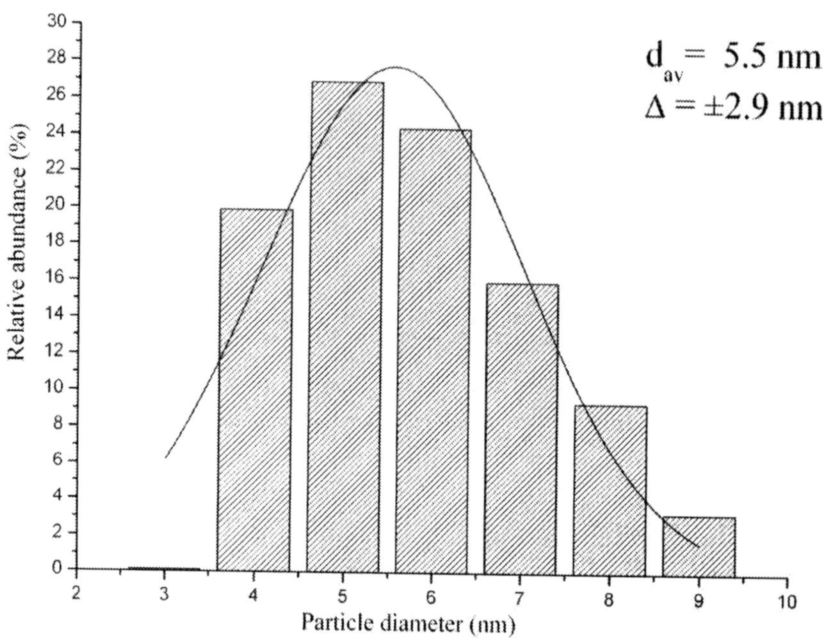

Figure 1. TEM micrograph and particle size distribution of silver nanoparticles in SNP4.

Transmission electron micrographs of water SNP solutions were made on LEO912 AB OMEGA microscope. Size distributions were calculated by the Gauss approximation for no less than 350 particles. The SNP1, SNP2 and SNP3 samples were obtained from the standard micellar solutions; particle diameters lied in the range 8-10 ± 4-6 nm; the corresponding electron micrographs and hystograms may be found elsewhere [17,29]. The nanoparticles in SNP2 and SNP3 samples were prepared in micellar solutions with slightly different water content, hence, within the size range mentioned above, it is assumed that in SNP2 the particles may be somewhat smaller than those in SNP3. The SNP4 and SNP5 solutions were prepared by the modified procedures, allowing to obtain more narrow distribution and (for SNP4) essentially smaller average size. The TEM images of these samples are presented in Figs. 1, 2. The nanoparticles are approximately spherical; particle sizes in SNP4 and SNP5 are 5.5±2.0 and 9.2±2.7 nm, respectively. In all five preparations electron diffraction patterns reveal the crystalline structure with parameters close to the gold standard. Measurements of zeta potential show that the outer surface of AOT shell bears negative charge resulting from the dissociation of anionic groups in polar headgroups of AOT molecules.

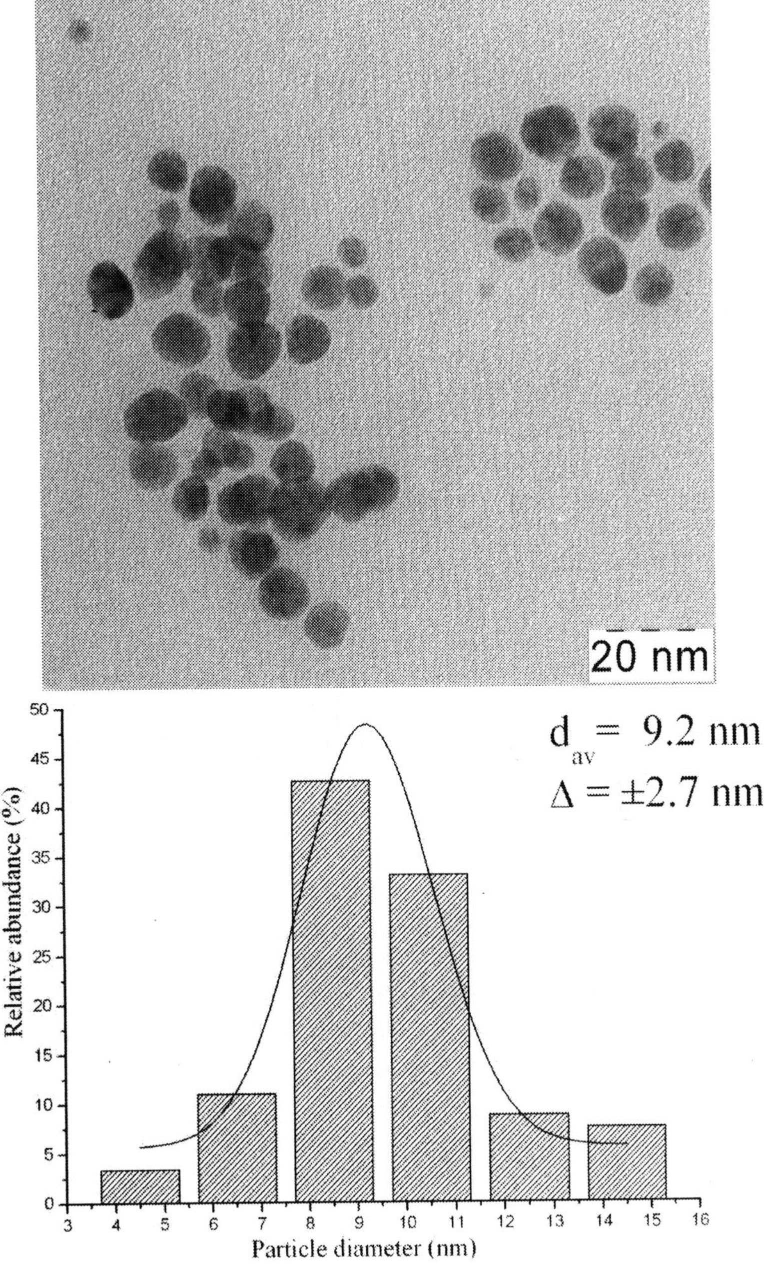

Figure 2. TEM micrograph and particle size distribution of silver nanoparticles in SNP5.

In experiments with *Physarum polycephalum* the effect of three chemical agents was studied, namely: SNP, AOT and Ag^+; the latter two were tested (either independently or as a mixture) in the concentrations equal to those added with the corresponding SNP solution. More precisely, the AOT solution was used as a control to the SNP solution, allowing us to distinguish between the effect of SNP themselves and that of SNP solution containing both SNP and AOT. Ag^+ ions or Ag^+ + AOT mixture were taken in order to elucidate whether the effect observed with SNP is comparable with action of Ag^+ ions.

Plasmodium of the *Physarum polycehallum* is capable of unlimited growth as a single multinuclear cell. The migrating plasmodium looks like a fan-shaped protoplasmic film, which is smooth and continuous at the front and transforms into a tree-shaped network of individual protoplasmic strands in more caudal regions (Fig.3). When submersed in liquid, the plasmodium disperses into microplasmodia, rounded multinuclear cells 100–200 μm in diameter. In macro- and microplasmodia, the protoplasm is differentiated into relatively stable ectoplasm and fluid endoplasm streaming through ectoplasm tubes (veins or strands) and channels penetrating into ectoplasm layer near the leading edge. The direction and the velocity of endoplasmic flow change in accordance with the pressure gradient produced by cyclic ectoplasm contraction-relaxation well coordinated along the plasmodium body [31]. Contractile activity of ectoplasm is provided by myosin oligomers interacting with actin filaments attached to the plasma membrane [32]. The period of contractile oscillations and shuttle streaming of the endoplasm varies in the range 1–5 min depending on the physiological state of the plasmodium [33] and external influence. Migration of plasmodium occurs due to the more intensive or more prolonged protoplasm streaming toward the leading edge. Any fragment of plasmodium restores the integrity of the plasma membrane [34] and resumes the contractile and motile activities. By this property, strands excised from a plasmodium can be used for force measurements, and film fragments of standard size and shape can be used in chemotactic assays.

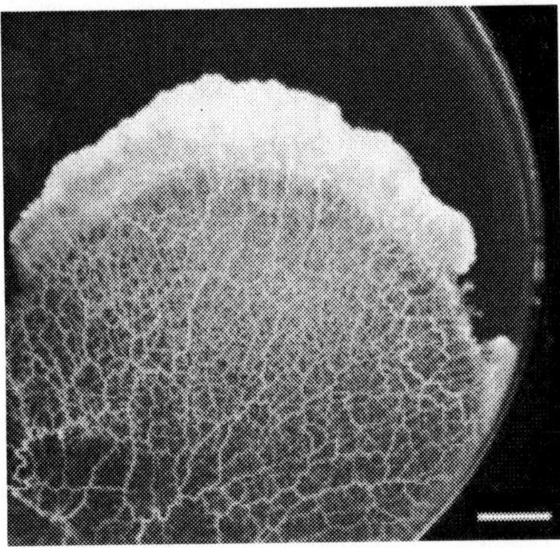

Figure 3. Spreading of *Physarum polycephalum* plasmodium on agar layer. 12 h after feeding. Scale bar, 1.5 cm.

Culture. Microplasmodia of *Physarum polycephalum* strain VKM F-3283 were grown in liquid medium [35] on a reciprocating shaker (Elban Laboratory Instruments, Poland) at 120 min–1.2 cm amplitude; the medium contained (g/l): peptone, 10; yeast extract, 1.5; glucose, 10; KH_2PO_4, 2; $CaCl_2 \times 2\ H_2O$, 0.6; $MgSO_4 \times 7\ H_2O$, 0.6; $FeCl_2\ \times 4H_2O$, 0.06; $MnCl_2 \times 4\ H_2O$, 0.084; $ZnSO_4 \times 7\ H_2O$, 0.034; hemin, 0.001 in 0.03 M citrate buffer, pH 4.6 (free calcium, 0.5 mM).

Macroplasmodia were obtained by fusion of microplasmodia and grown on 1.5 % agar, with oat flake feed [36]. Five to six hours after feeding, the plasmodium began to spread over the substrate, and as the leading edge protruded, the continuous protoplasmic sheet transformed into a network of interconnected strands (Fig.3). After 12–15 h, the strand segments and film fragments were isolated to use in experiments.

Viability tests. Microplasmodia were precipitated from the culture medium for 1 min at 600*g*, washed with distilled water containing 0.1 mM $CaCl_2$, pelleted and resuspended in fresh culture or salt medium. Tested agents were added and the portions of microplasmodial suspension (10 ml) were shaken for 1 h in conical flasks. The necrotic death was evidenced by the release of yellow plasmodial pigment. Viability of the protoplasmic strands was estimated by the release of the pigment and resumption of motility upon transfer onto agar substrate.

Chemotactic tests. The response to a chemical agent in solution was assayed by the change in the mode of auto-oscillations of the longitudinal force generated by an isolated strand under isometric conditions. The force was registered by highly sensitive tensiometer [37] equipped with an electromechanical transducer permitting controlled extension of the strand (Fig.4). A strand segment isolated from macroplasmodium, 0.3–0.5 mm in diameter and about 4 mm long, was fixed horizontally to the tensiometer hooks with agar and immersed into the glass chamber with control solution. To prevent precipitation of silver as AgCl, in the standard control solution (HEPES, 10 mM, pH 7.0; $CaCl_2$, 0.1 mM [38]) $CaCl_2$ used to stabilize the physiological state of the plasmodium, was replaced with $Ca(NO_3)_2$.

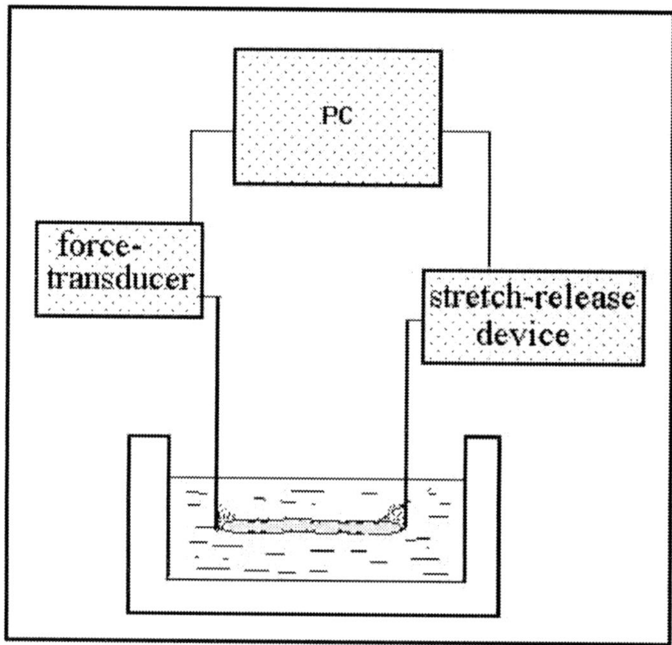

Figure 4. Scheme of equipment used for measurement of the force generated by the plasmodial strand under isometric conditions.

Elastic properties of the ectoplasm were estimated by the amplitude of the fast phase of force response to the abrupt strand extension [37].

The response to a chemical agent introduced into substrate in spatially uniform concentration (further referred to as a temporal test) was assayed by the area of spreading of standard plasmodium samples over a 2–3-mm layer of 2% agar in a Petri dish. Two types of the standard samples (4 mm diameter), were cut out of the smooth frontal zone of a macroplasmodium either with an underlying agar layer (first type) or with the supporting disk of filter paper (second type). In the latter case, the disks were placed in front of the leading edge, and plasmodium was allowed to cover them when propagated.

For direct comparison of the repellent action in spatial assay the modified “double strip” test was used [23]. Each standard sample was placed at the interface between agar strips separated by 1-mm gap and containing different agents (alone or their combinations). Experiments were performed in plastic boxes each containing 6 of such paired strips or in five-section chambers. A gap or a wall separating the paired strips permitted to avoid a coalescence of plasmodia. All samples in each experiment were excised from one macroplasmodium.

Aliquots of concentrated water solutions of SNP (1-5), $AgNO_3$ and AOT were thoroughly mixed into 10 mM HEPES (pH 7.0)-containing melted 2% agar (Difco, USA) cooled to 50°C.

In all experiments, the final concentrations of Ag^+ and AOT were equal to the silver and AOT concentrations obtained upon introducing the corresponding SNP solution.

Experiments were carried out at 20°C, experimental chambers were protected from light in view of the photosensitivity of both plasmodia [39] and silver compounds. Images of plasmodia were recorded in digital form with UMAX Astra 6700 (UK) scanner.

Results and Discussion

Our studies may be divided into two main parts: experiments in water solution and on agar substrate. In water solution we made the lethality tests and determined the effects on the contractile auto-oscillations at various nanoparticle concentrations. The results were compared with those obtained on the other biological objects (1) with similar nanoparticle preparations and (2) with silver nanoparticles of different origin from the data reported in literature.

On the agar substrate we estimated the biocidal effects by repellent activity both in temporal (decrease in area of spreading) and spatial tests (directional motile response). Experiments in water solution were made with SNP1 preparation. On agar substrate, comparison of the repellent efficiency of SNP, AOT and Ag^+ ions were fulfilled also with SNP1; for the comparison of the repellent activity of different SNP preparations in the spatial tests we used first SNP2 and SNP3, then SNP4 and SNP5.

I. Experiments in Water Solution

Viability of Plasmodium and Suppression of Contractile Auto-Oscillations

First, we determined the concentrations of SNP lethal for microplasmodia in the standard growth medium [35], in its salt–buffer base, and in control solution (HEPES, 10 mM, pH 7.0; $Ca(NO_3)_2$, 0.1 mM). Upon 1h incubation in both salt media, SNP1 diluted to the concentration 10 μg-ion/l (1.08 μg/ml) of silver caused membrane disintegration and death of microplasmodia, evident by decoloration of cells and release of the yellow pigment into solution. This destruction was slowed down in the growth medium, presumably because of the presence of peptone and yeast extract, which can bind with the nanoparticles. When incubated with 1 μg-ion/l of SNP1, microplasmodia fully preserved membrane integrity and motility. Longer incubation in the salt medium containing 1 μg-ion/l of SNP1 caused cyst formation normal for starving microplasmodia [40].

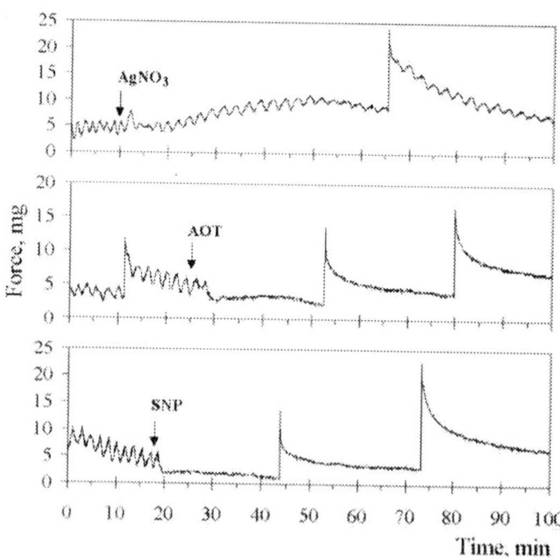

Figure 5. Effect of SNP1 and equivalent amounts of $AgNO_3$ and AOT on the auto-oscillations of isometric force. C($AgNO_3$)=100 μM; C(AOT)=560 μM; C(SNP)=100 μg-ion/l. Vertical shifts in the right-hand part mark the additional stretching. From [18].

Second, we registered the oscillations of force generated by isolated strands of macroplasmodium in water solution. As can be seen in Fig. 5, the oscillations ceased in few minutes upon addition of either SNP1 at 100 μg-ion/l (10.8 μg/ml) or equivalent AOT concentration. Upon addition of an equivalent amount of $AgNO_3$, after a brief break the oscillations continued with a doubled period and increasing force (oscillation damping in this experiment was observed only after 3 h). It is also evident from the figure that cessation of oscillations in the presence of SNP or AOT cannot be explained by strand 'relaxation' (i.e., loss of tension), because oscillations were not restored upon stepwise 20% stretch (vertical shifts in the recordings). The force response to stretch testifies to the preservation of elastic properties of the plasmodium strand [37].

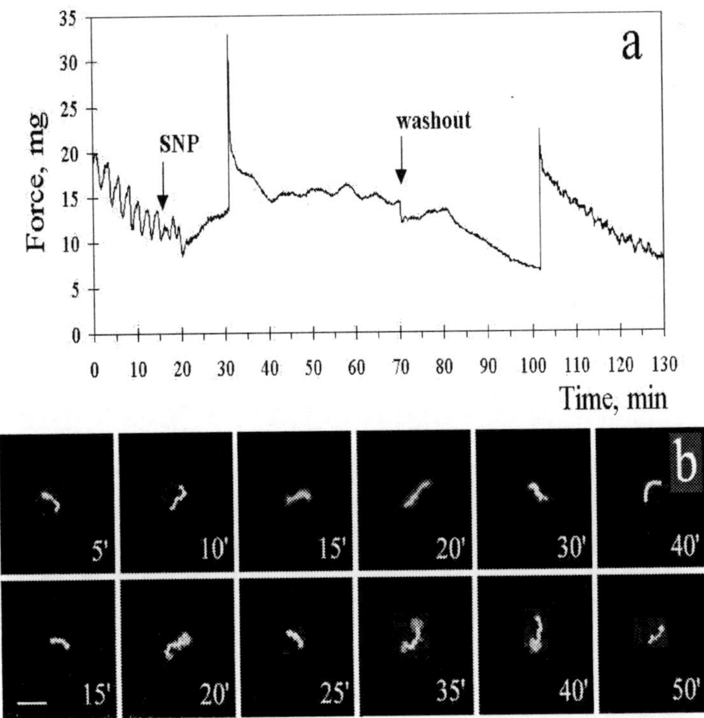

Figure 6. Effects of SNP on the viability of plasmodium: dependence on the concentration and time of incubation. (a) Reversibility of the auto-oscillation stoppage induced by 30 μg-ion/l of SNP1. (b) Viability of the plasmodium under action of 100 (upper row) and 30 μg-ion/l (lower row) of SNP1. Timing marks indicate the term of incubation in SNP1 solution; images made 4h after the washout and placing the strands on agar layer are shown. Scale bar, 1 cm.

In the presence of 100 μg-ion/l SNP1 (Fig. 6b, upper row) or equivalent AOT concentration, the strands lost their pigment in 10-15 min and became unable of substrate attachment and migration. With $AgNO_3$ these signs of death were observed much later, after 1.5–3 h depending on the state of culture (not shown).

At the nanoparticles concentration lowered to 30 μg-ion/l (3.24 μg/ml), contractile oscillations also ceased during first 10-20 min after SNP1 addition. In response to the strand stretching, oscillations temporally recommenced with a markedly increased period and ceased again when the tension was released. The effect is not due to plasmodium death since after SNP washout the strands resumed contractile activity (Fig. 6a). As seen from Fig. 6b, the strands remained viable during approximately one hour incubation in the SNP-containing control solution, no signs of the loss of pigment or cell fragmentation characteristic for apoptosis were observed. Moreover, 4 h after the SNP washout the strand segments began to spread over the agar layer.

Reversibility of the SNP effect suggests that the targets essential for survival, first of all mitochondria, remain intact. Though the latter does not exclude SNP-induced changes in their functional activity, such changes should be reversible and unable to provoke the cell death. Identification of the cell membrane SNP targets responsible for the suppression of contractile auto-oscillations obviously requires special investigation, which is beyond the scope of this study.

The experiments described in this section have demonstrated also that at lethal concentration such characteristic as the suppression of contractile auto-oscillations can not reveal the difference in biocidal activity between SNP and AOT.

Sustained Auto-Oscillations as a Measure of Biocidal Efficiency

When the concentrations of all tested agents were lowered by one order of magnitude (to 10 μg-ion/l or 1.08 μg/ml), as compared with those lethal for SNP and AOT, the strands remained viable permitting a long-continued measurement of the contractile auto-oscillations. Recording the isometric force (Fig. 7) demonstrated the increase in the oscillation period characteristic of repellents [22, 25, 26]. The effect differed both in magnitude and time-course depending on the agent tested. For the quantitative estimation of chemotactic response, we determined the change in the period relative to its initial value.

By this index, SNP proved to be 1.5, 2.5, and 5 times more effective than $AgNO_3$+AOT, AOT, and $AgNO_3$, respectively.

Thus, such an assay is quite discriminative; besides, it allows quantitative assessment and monitoring of the dynamics of chemotactic response.

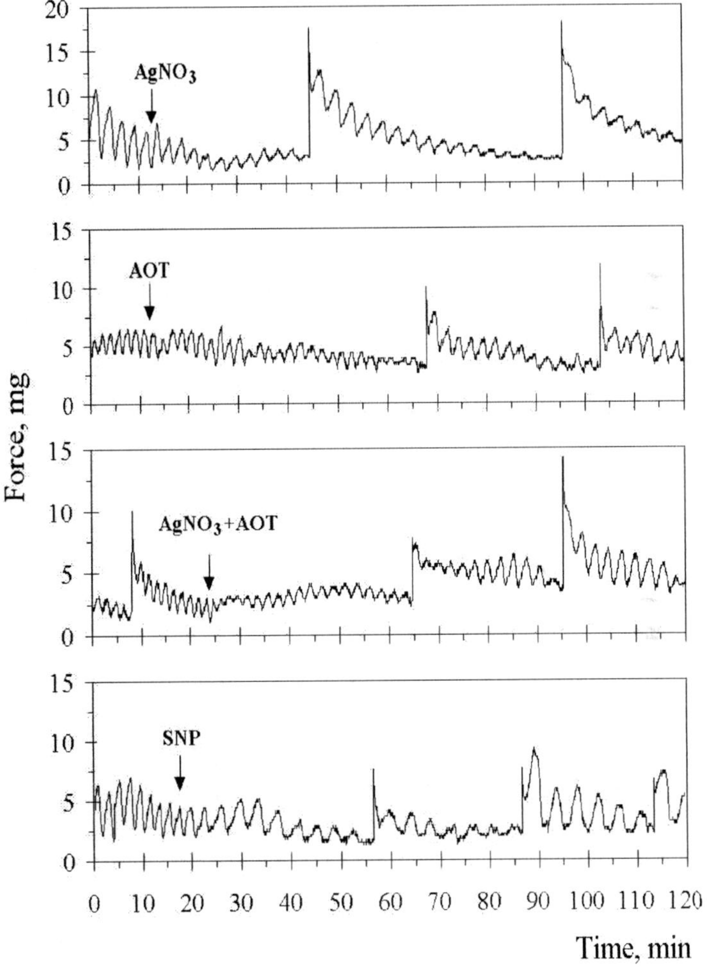

Figure 7. Effect of SNP1 and equivalent amounts of $AgNO_3$ and AOT on the increase in auto-oscillation period. Chemotactic efficiency was calculated as $[(T_i - T_0) / T_0] \times 100\%$, where T_0 and T_i are the periods (averaged for 10 oscillations) in the control and in the presence of the agent; for $AgNO_3$, AOT, $AgNO_3$+AOT, and SNP this index was equal to 23, 40, 58, and 100%, respectively. C($AgNO_3$)=10 μM; C(AOT)=56 μM; C(SNP)=10 μg-ion/l. From [18].

It seems useful to compare the characteristic lethal and sublethal concentrations of SNP for the plasmodia obtained in our work with those found in studies of the toxicity of silver nanoparticles towards bacteria cells and mammalian cultured cells in water solution. For the comparison we take only the results obtained for nanoparticles of the same size range (below 20 nm) prepared by the biochemical synthesis (for bacteria cells [17]) and by the other techniques [8,11,41]. The data are summarized in Table 1.

Table 1. Toxic effects of silver nanoparticles observed on various biological objects in water solution

Object	Lethal SNP concentrations (μg/ml)	Sublethal SNP concentrations (μg/ml)	Source
Microplasmodia	1.08 necrosis	0.108 < C(SNP) < 1.08	[18] and this work
Macroplasmodia	10.8 necrosis	1.08(**) ≤ C(SNP) ≤ 3.24(***)	[18] and this work
Bacterial cells (*) E.coli S. aureus	5, 2.88 5.4 2.88		[17] [8] [17]
Spermatogonial stem cells	8.75 (MTT EC_{50}) Drastic reduction of mitochondrial function and cell viability	2.50 (LDH EC_{50}) Slight increase in LDH leakage, cell apoptosis rather than necrosis	[11]
HT-1080 A431 cells	12.5 necrosis 12.5	6.25 Oxidative stress 0.78 apoptosis 6.25 Oxidative stress 1.56 apoptosis	[41]

(*) 100 % death after 30 min exposure
(**) Two-fold increase in the period of contractile oscillations
(***) Reversible suppression of the contractile oscillations

By sublethal concentrations we mean here those provoking suppression of the life functions, but not leading to necrotic cell death. It is seen that lethal concentration found for macroplasmodium is close to those obtained for mammalian cultured cells [41] and somewhat higher than those measured for bacteria cells [8,17]. However, in the latter case the difference may prove to be smaller after testing for macroplasmodium the SNP concentrations lower than

10.8 μg/ml (but higher than 3.24 μg/ml). At the same time, this testing may result in the widening of the range of sublethal concentrations. Anyhow, judging from the present data on lethal concentrations, *Physarum* macroplasmodium seems to be approximately as sensitive to SNP as bacterial and mammalian cells. For microplasmodia, the lethal SNP concentrations are an order of magnitude lower. This may be mainly due to the increase in the plasma membrane area to cell volume ratio taking place upon transfer from macroplasmodia to microplasmodia. It is also probable that, for microplasmodium, the *Physarum* capacity to repair plasma membrane dam-ages [34] is limited due to much smaller cell volume.

As for the sublethal concentrations, it is seen that, for both types of mammalian cells presented in the table, they strongly depend on the type of assay used for the evaluation of SNP toxicity. For spermatogonial stem cells, the test on lactat dehydrogenase leakage occurring due to disrupting the plasma membrane gives the SNP concentration within the sublethal range found for macroplasmodium, while the test on mitochondrial function gives the SNP concentration several times higher and close to that lethal for macroplasmodium. For HT-1080 and A431 cells oxidative stress and the increased lipid peroxidation took place at SNP concentrations much higher than those which stimulated the beginning of apoptosis. So it is clear that, for the correct comparison of the SNP toxicity on various biological objects at sublethal concentrations, it is necessary to compare the results obtained in one and the same type of assay. Still it may be noted that, for HT-1080 and A431 cells, the SNP concentrations required for the onset of apoptosis are close by order to the sublethal range found for microplasmodia, the fact that probably indicates to the similar origin of the SNP toxic action.

II. Experiments on Agar Substrate

Assessment of the Relative Efficiency of Agents Introduced Into Agar Substrate with the Use of Temporal and Spatial Chemotactic Tests

The organisms, like *Physarum* or *Dictyostelium*, in contrast to animal cells grown in surface culture, give an opportunity to mimic one-sided contact of cell with biocides-containing materials (for example, with gauze pad applied to a skin surface).

When the agents at high concentrations (100 μg-ion/l SNP1 and equivalents) were introduced into the agar substrate, plasmodia remained motile for more than 36 h; longer incubations in all cases eventually caused death and leakage of the pigment. Thus, the biocide activity in the support layer is markedly decreased as compared to the water solution. For SNP1 and AOT, this may be due to the binding with agar and, hence, lowering of their activity in this medium. It should be also borne in mind that, when plasmodium is on a solid substrate, the area of plasma membrane contact with the agent is less than half of that in solution. Further, surface growth and propagation are natural conditions for the plasmodium, and its protective mechanisms here can work more efficiently than in water solution.

In all experiments when the substrate contained SNP1 or AOT, the starving plasmodium lost more mass than in control; this was especially pronounced with SNP1, and was indirectly indicative of the operation of a protective mechanism requiring energy consumption.

When the concentration of agents in the agar was lowered by one order of magnitude, normal starvation-induced encystment [40] was observed on the second day in all cases except for SNP. The SNP toxic action was expressed in the change of plasmodium color, less regular protoplasm streaming and appearance of local protoplasmic clots.

Though the lethal effect of the agents in agar substrate was decreased, their repellent action was well pronounced. At concentrations equivalent to 10 μg-ion/l, the area of plasmodium spreading after 3 h was smaller than in the control. At tenfold higher concentration, spreading was completely suppressed in all cases except for $AgNO_3$. Another characteristic feature of repellent action was the accelerated transition from spreading to migration [26]. Figure 8 displays the results of this assay 6 h after the beginning of experiment. One can clearly see the reduction in area of spreading at higher concentrations (lower row). It is also evident that at lower concentrations (upper row) plasmodium on substrate with SNP1 is in the migrating form and occupies much less surface than with $AgNO_3$, AOT, or $AgNO_3$+AOT.

Thus, assessment of the lethal and repellent effects of the tested agents in solution and in solid substrate yield quite similar results regarding their comparative efficacy, and demonstrate the expedience of testing at sublethal concentrations.

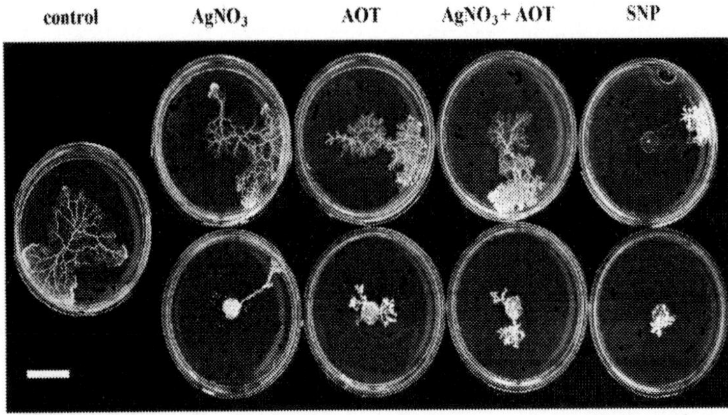

Figure 8. Spreading of plasmodia on agar substrate containing the tested agents 6 h after the beginning of assay. Upper row, concentrations as in Fig. 7; lower row, ten times higher. Scale bar, 1 cm. From [18].

Nonetheless, the resolution of the temporal testing proved insufficient to reliably reveal the difference in spreading for AOT and $AgNO_3$+AOT. Since the possibility of discerning small differences in biological activity may prove to be useful in evaluating the effects of SNP parameters (size, form or surface charge), we examined the capacity of spatial tests, which, as a rule, are superior in this respect [24, 26].

Testing was performed on paired agar strips with agents introduced at concentrations equivalent to 10 μg-ion/l of SNP. Figure 9 displays the comparison of SNP1 vs. $AgNO_3$ +AOT; the direction of taxis clearly demonstrates that the repellent activity of SNP exceeds the sum activity of the other two agents.

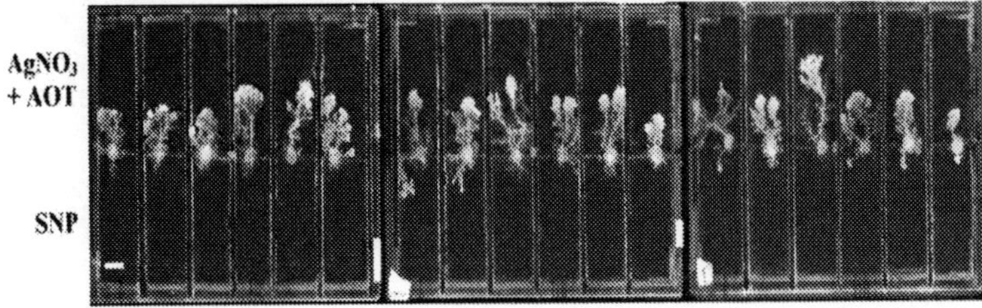

Figure 9. Spatial test for comparison of the repellent action (negative chemotaxis of plasmodia away from the stronger agent). C(SNP1)=10 μg-ion/l; C($AgNO_3$)=10 μM; C(AOT)=56 μM; 10 h after the plasmodia were placed at the interface between strips. Scale bar, 1 cm. From [18].

The same testing was done for $AgNO_3$ vs. $AgNO_3$+AOT and AOT vs. $AgNO_3$+AOT. Negative taxis away from AOT in the presence of $AgNO_3$ in both strips was observed on all 18 samples; away from $AgNO_3$ with AOT in both strips, on 13 samples out of 18. The high discriminative ability of this spatial test confirmed and refined the data obtained in temporal tests; the order of relative efficiency is $AgNO_3 \ll AOT < AgNO_3+AOT \ll SNP$.

Thus, the chemotactic behavior of *Physarum* plasmodia shows that in nanoparticles obtained by the biochemical synthesis the known biocidal properties of metallic silver are greatly enhanced, and this effect cannot be reduced to the membrane destabilizing action of the surfactant shell surrounding a nanoparticle.

These results are in line with the noticeably higher toxicity of SNP vs. Ag+ ions observed in water solutions both with biochemically synthesized nanoparticles against bacteria *E. coli* [17] and with silver nanoparticles obtained by the other methods against *E. coli* [6,9], mammalian cells [11] and fish embryos [13]. It is clear also that SNP as a biocide are significantly more active than AOT molecules. The results obtained with Ag^+ or Ag^+ + AOT mixture exclude the possibility to explain the SNP effect by the action of Ag^+ ions released from the SNP surface, as suggested by some other authors [8].

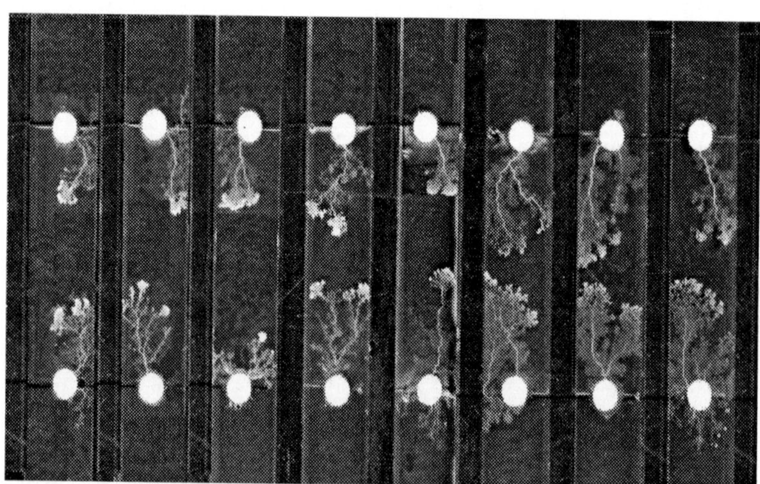

Figure 10. Threshold plasmodium reaction observed under spatial testing vs. control solution (in the center of the chamber) at C(SNP1)=3 μg-ion/l. 6h after the samples (with supporting disks of filter paper) were placed at the interfaces between the SNP- and the control solution-containing strips. Scale bar, 1 cm.

To complete our analysis of the biological effects of SNP1, we determined the threshold concentration for the negative taxis. The SNP concentration provoking 0.25±0.17 temporal increase in the oscillation period (MV± SD; n 7) was found equal to 3 μg-ion/l. Figure 10 shows the results of spatial testing done for 3 μg-ion/l SPN1 vs. control solution, the control solution-contained strips are in the central part of the chamber. It is seen that, in spite of the decreased effectiveness of SNP in the agar substrate, all tested samples exhibited negative taxis from SNP. Thus, a comparison of the results obtained in both assays proves the advantage of spatial testing and high sensitivity of chemotaxis itself.

Inasmuch as chemotaxis is central event in immunity, wound healing and repairing the tissue damage, the concentration threshold of avoidance reaction is essential for proper clinical application of the nanoparticles. As a first approach, the results obtained with *Physarum* plasmodium should obviously be taken into consideration, because the involvement of such important signaling pathways as PI3K/PTEN in the plasmodium chemotaxis [42] suggests its common nature with that of mammalian cells.

Comparison of the Repellent Action for Different SNP Preparations

The goal of further investigation was to elucidate whether chemotactic tests are sensitive enough to reveal the difference in biological efficiency of various SNP preparations. To make this we used the spatial test with high resolution described above. First, we compared the samples SNP2 and SNP3 with sizes presumably different within the whole range in our polydisperse preparations (3-16 nm, see Experimental section).

When introduced into melted agar, both samples were diluted to the same final silver concentration of 10 μg-ion/l. The final AOT concentrations in agar for SNP2 and SNP3 were 90 μM and 39 μM, respectively.

The plasmodium samples were placed on the interface between SNP2 and SNP3-containing agar strips, and plasmodia were allowed to avoid more strong repellent. As seen from Fig. 11a, at 3h after the beginning of experiment, most of the plasmodia displayed bidirectional spreading with some preference to the SNP3-contained strip (to the right). In the course of time, the frontal zones on SNP2-contained strips reduced, the plasmodia became oriented and exhibited negative taxis from SNP2.

In another variant of testing, 50μM AOT was added to the SNP3-contained agar to equalize its content in both strips. The experiments with and without AOT addition were made simultaneously on the samples excised from one and the same plasmodium. For quantitative estimate of the results, we used chemotactic index (Cht Index), a ratio of number of the fronts oriented towards less effective repellent to their total number. Mean values of this index ±SD obtained in 7 parallel experiments (with 6-10 plasmodium samples for each variant) are shown in Fig. 11b. As seen from the figure, addition of 50μM AOT did not reduce the value of Cht Index. So, it is clear that the initial difference in AOT concentration in SNP2- and SNP3-containing agar can not be the reason of avoidance reaction, and the directional response is connected with the difference in biocidal and, hence, repellent action of nanoparticles themselves. Taking into consideration the equality, in SNP2 and SNP3 samples, both of the nanoparticles concentration and surface charge of their protective shell, it is reasonable to suggest that the negative taxis from SNP2 is conditioned by the difference in particle sizes. As follows from some reports on the biological activity of silver nanoparticles [8,9], smaller particles are more toxic than larger ones. Therefore, the avoidance of SNP2 may be the result of stronger toxicity of the smaller particles which are supposed to prevail in this sample.

To check for this possibility, we tested the two samples with obviously different average particle sizes and more narrow size distributions – SNP4 and SNP5 (see Figs.1 and 2).

Testing was made on the agar strips containing 3 μg-ion/l of SNP4 and SNP5. At this SNP concentration the difference in AOT content between the two agar gels is 24μM, hence, in view of the results described above, possible contribution of free AOT in chemotatic response can be excluded. The result of testing is shown in Fig.12. As one can see, all plasmodia are oriented towards SNP5 (in the center), and, consequently, SNP4 is more strong repellent. This confirms the supposition about the larger toxicity of smaller nanoparticles. It seems probable that the effect of particle sizes is connected mainly with the difference in their number per unit volume: for a given overall concentration of silver, the number of smaller nanoparticles will exceed that of the nanoparticles of larger size. In our case, the difference in the particles number per unit volume is about one order of magnitude: SNP4 - 1.48×10^{14}; SNP5- 3.16×10^{13} (particles/ml).

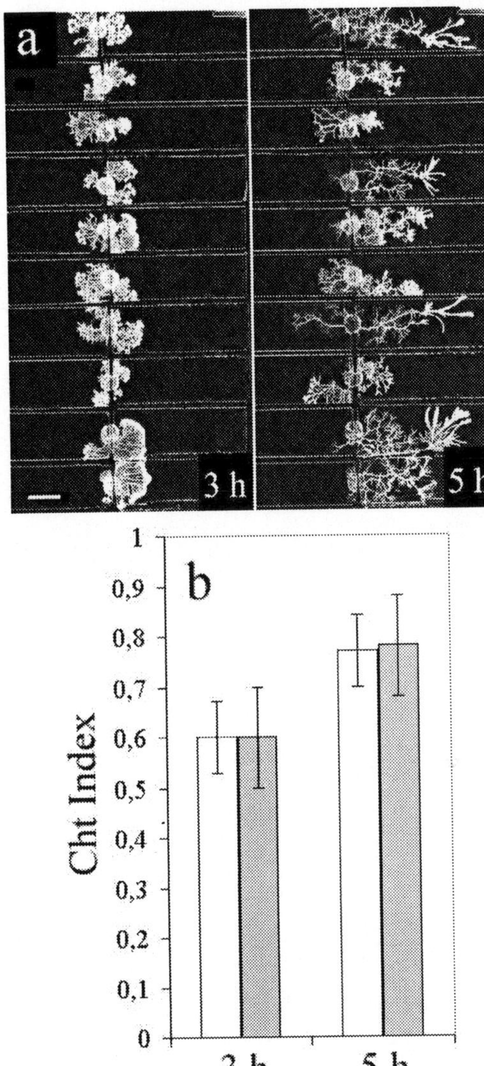

Figure 11. Comparison of the repellent action of SNP2 and SNP3 at the equal concentration of 10 μg-ion/l. (a) Successive images showing the development of negative taxis from SNP2 (on the left-hand side of the chamber). Scale bar, 1 cm. (b) The time-dependent increase in ChT index (calculated as a portion of plasmodial fronts oriented towards less effective repellent) at direct comparison between SNP2 and SNP3 (white columns), and after the addition of AOT to SNP3-containing agar (gray columns). Mean values ± SD obtained in 7 experiments are shown.

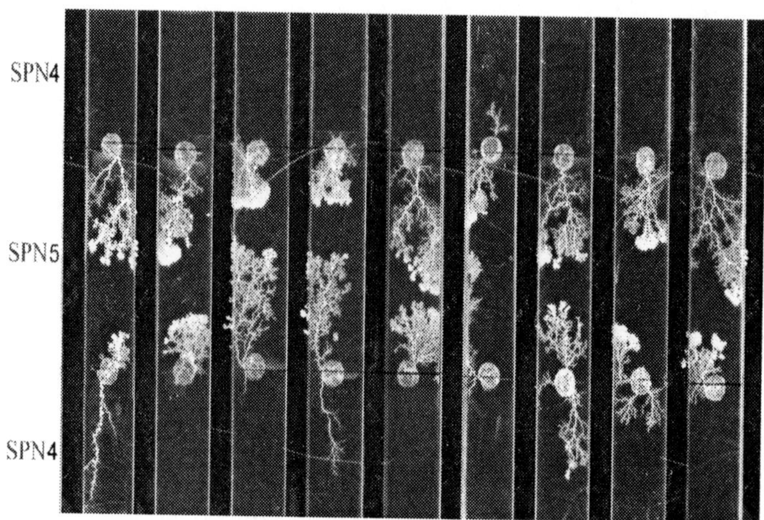

Figure 12. Comparison of the repellent action of SNP4 and SNP5 at the equal concentration of 3 μg-ion/l. Directional movement of all plasmodia towards SNP5-containing strips (in the center of the chamber) proves higher repellent efficiency of SNP4. Scale bar, 1 cm.

Conclusion

Biological activity of silver nanoparticles obtained by the original method of biochemical synthesis was tested on plasmodium of the acellular slime mold *Physarum polycephalum*, a multinuclear amoeboid cell with auto-oscillatory mode of locomotion. Experiments were carried out on micro- and macroplasmodia both in water solution and on the agar substrate, at various nanoparticle concentrations. Also, the paired tests were made to compare the effect of different nanoparticle preparations. In water solutions, the effect of silver nanoparticles on the viability of *Physarum* was determined by registration of the necrotic death revealed by the loss of pigment (for micro- and macroplasmodia) and by measuring changes in the contractile oscillation period of isolated strands (for macroplasmodia). The lethal nanoparticles concentration for microplasmodia was found to be an order of magnitude lower than that for macroplasmodia (1.08 and 10.8 μg/ml, respectively). As one of the possible causes, a decrease in the plasma membrane surface to cell volume ratio may be considered taking place upon the transfer from micro- to macroplasmodia, this decrease leading to the reduction of the area of effective cell interaction with nanoparticles.

At sublethal nanoparticle concentrations it was found possible (1) to estimate the difference in efficiency of the toxic action between the nanoparticles and stabilizing surfactant (AOT) present in the nanoparticles solution and (2) to compare the effect of nanoparticles with that of the equivalent concentration of Ag^+ ions. It was established that the toxicity of nanoparticles exceeds that of both AOT and silver ions. Deserves attention our observation made at 3.24 μg/ml SNP, on the reversibility of the suppression of contractile oscillation and hence, occurrence of a concentration range, where a cell remains alive despite the inhibition of its functional activity, the fact which suggests the existence of a certain defense mechanism(s). Whether we deal with an extrusion of the nanoparticles penetrated through the plasma membrane or just with *de novo* synthesis of damaged lipids and/or proteins present therein, remains unclear. In both cases, at a given area of plasma membrane, the cell volume and, probably, additional factors determining a reparative capacity of the cell become decisive.

Studies on the agar substrate showed that toxic action of the nanoparticles in the support agar layer was markedly weakened in comparison with that in water solution. In view of the known chemical activity of metal nanoparticles and their tendency to adsorption on various surfaces, a reasonable explanation of this effect may be a lowering of their effective concentration due to the binding with the agar polymer structures. If so, it is possible to assume that biocide activity of the nanoparticles may be attenuated by the other polysaccharides, including extracellular matrixes of mammalian tissues.

Judging from the changes in the area of spreading and from negative chemotaxis on the paired strips, it was deduced that relative effectiveness of the agents tested changed in the row: $AgNO_3 \ll AOT < AgNO_3 + AOT \ll$ SNP. This sequence supplements and refines the results obtained on macroplasmodia in water solutions. The higher efficiency of silver nanoparticles compared to that of silver ions agrees also with the data available from literature, concerning the observations made in studies of the nanoparticles interaction with bacteria cells [6,7,9,17], mammalian cultured cells [11,12] and fish embryos [13]. This may be regarded as the argument in favor of the opinion, that biological action of silver nanoparticles can not be reduced to the effects of silver ions, but implies another mechanism where the nanoparticles themselves play an important role.

Additional indication to the role of nanoparticles comes from our experiments on the paired agar strips with preparations containing nanoparticles of different sizes: it turned out that smaller (5 nm) nanoparticles were more toxic than the larger ones (9 nm), in accordance with the results obtained for silver nanoparticles of various sizes at their interaction with bacteria cells [8,9]. This shows probably to the significance of numerical concentration of nanoparticles per unit volume of the medium: at the equal silver concentration, numerical concentration of the smaller particles will be higher, thus providing the possibility of the more intensive adsorption onto the cell surface.

In summary, our study of the biological effects of silver nanoparticles with the use of chemotactic-based assays and *Physarum* plasmodium as a test-organism, permitted to prove that chemotactic tests, as a tool, provide a number of essential advantages. Made at concentrations considerably below those used for lethality assays, they are able to distinguish between the effectiveness of the two strong biocides (like SNP and AOT in our study). The lowering of concentration as an expedient enabled us to increase the discriminative capacity of all assays suggested here: the monitoring of plasmodium spreading on biocide-containing substrates, the directional movement in the paired experiments, and the contractile activity measurements. The latter two assays permit also the control of dynamics and quantitative analysis of cell response. There are grounds to believe that, due to the unique sensitivity of chemotaxis itself, the instrumental and behavioral tests applied may prove to be more sensitive than those based on registration of any other type of a cell activity. The results obtained on *Physarum* plasmodium allow to recommend a similar approach for differential testing and revealing trace amounts of harmful substances with the use of any pro- and eukaryotic cell capable of chemotaxis.

With respect to the toxic properties of silver nanoparticles it is confirmed that, as a biocidal agent, they are superior both to silver ions and to the surfactant molecules present in the nanoparticles preparation. It is shown that the biocidal activity of silver nanoparticles manifests itself not only with prokariotic, but also with low-eukariotic cells, like *Physarum* plasmodium. This makes these nanoparticles promising both as a bactericide and as a fungicide, though potentially dangerous for mammalian cells. Hence, application of the biocidal properties of silver nanoparticles may give useful devises in therapy, but it should be made with caution because of the risk of tissue damage during treatment. Another aspect, which must be taken into account, is the direct involvement of chemotaxis in immunity and tissue reparation. Since the mechanism of chemotaxis is highly conserved between mammals and slime molds [43, 44], low SNP concentration threshold for the avoidance reaction found in our work suggests that the directional movement of mammalian cells towards the gradient of bacterial peptides, cytokines and growth factors could be substantially impaired by the contradictory action of silver nanoparticles. Therefore, in elaboration of a protocol for clinical application of silver nanoparticles, it is necessary to include investigations of the chemotactic behavior of tissue cells.

Acknowledgments

This work was supported by the Russian Foundation for Basic Research, grant N 04-04-49095 .

References

[1] Salata, O.V. Applications of nanoparticles in biology and medicine. *Journal of nanobiotechnology*, 2004, 2:3 doi:10.1186/1477-3155-2-3.

[2] Stratmeyer, M.E.; Goering, P.L; Hitchins, V.M. et al. What we know and do not know about the bioeffects of nanoparticles: developing experimental approaches for safety assessment. *Biomed. Microdevices,* 2008, DOI 10.1007/s10544-008-9261-9.

[3] Hoet, P.H.M.; Bruske-Hochlfeld, I; Salata, O.V. Nanoparticles – known and unknown health risk. *Journal of nanobiotechnology*, 2004, v.2, p.12.

[4] Oberdorster, G.; Maynard, A.; Donaldson, K. et al. Principles for characterizing the potential human health effects from exposure to nanomaterials: elements of a screening strategy. *Particle and Fibre Toxicology,* 2005, v.2, no 8.

[5] Mossman, B.T.; Borm, P.J.; Castranova, V. et al. Mechanisms of action of inhaled fibers, particles and nanoparticles in lung and cardiovascular deseases. *Particle and Fibre Toxicology,* 2007 May 30, available from: http//www.particleandfibretoxicology.com/content/4/1/4.

[6] Sondi, I.; Salopek-Sondi, B. Silver nanoparticles as antimicrobial agent: a case study on E,coli as a model for Gram-negative bacteria. *J. Colloid Interface Sci.*, 2004, v.275, p.177-185.

[7] Williams, D.N.; Ehrman, S.H.; Holoman, P. Evaluation of the microbial growth response to inorganic nanoparticles. *Journal of nanobiotechnology*, 2006, 4:3 doi:10.1186/1477-3155-4-3.

[8] Lok, C.-N.; Ho, C.-M.; Chen, R. et al. Silver nanoparticles: partial oxidation and antibacterial activities. *J.Biol.Inorg.Chem.*, 2007, v.12, p.527-534.

[9] Neal, A.L. What can be inferred from bacterium-nanoparticle interactions about the potential consequences of environmental exposure to nanoparticles? *Ecotoxicology*, 2008, v.17, p.362-371.

[10] Elechiguerra, J.L.; Burt, J.L.; Morones, J.R. et al. Interaction of silver nanoparticles with HIV-I. *Journal of nanobiotechnology*, 2005, 3:6 doi:10.1186/1477-3155-3-6.

[11] Braydich-Stolle, L.; Hussain, S.; Schlager, J.J.; Hofmann, M.-C. In vitro cytotoxicity of nanoparticles in mammalian germline stem cells. *Toxicological sciences,* 2005, v.88, p.412-419.

[12] Hussain, S.; Hess, K.; Gearhart, J. et al. In vitro toxicity of nanoparticles in BRL3A rat liver cells. *Toxicological sciences*, 2006, v.92, p.456-462.

[13] Asharani, P.V.; Wu, Y.L.; Gong, Z.; Valivaveettil, S. Toxicity of silver nanoparticles in zebrafish models. *Nanotechnology*. 2008, v.19, p.255102.

[14] Ji, J.H.; Jung, J.H.; Kim, S.S. et al. A twenty-eight-day inhalation toxicity study of silver nanoparticles in Sprague-Dawley rats. *Inhalation toxicology*, 2008, v.19, p.857-871.

[15] Kim, Y.S., Kim, J.S.; Cho, H.S. et al. Twenty-eight-day oral toxicity, genotoxicity and gender-related tissue distribution of silver nanoparticles in Sprague-Dawley rats. *Inhalation toxicology*, 2008, v.20, p.575-583.

[16] Pal, S.; Tak, Y.K.; Song, J.M. Does the antibacterial activity of silver nanoparticles depend upon the shape of the nanoparticles? *Applied and Environmental Microbiology*. 2007, v.73, p.1712.

[17] Egorova, E.M. Biological effects of silver nanoparticles. In: (Editor) *"Silver nanoparticles: Properties, Characterization and Applications"*. Nova Science Publishers, INC, 2010, New York, USA. (in press).

[18] Matveeva, N. B.; Egorova, E. M.; Beilina, S. I.; Lednev, V. V. Chemotactic Assay for Biological Effects of Silver Nanoparticles. *Biophysics*, 2006, v.51, p.758–763.

[19] Matveeva, N. B., Egorova, E. M.; Beylina S. I. Chemotactic assay is capable to reveal the difference in efficiency of nanosized silver particles. In Z. A. Podlubnaya & S. L. Malyshev (Eds.), Biological Motility: Achievements and Perspectives, 2008, v.2, p.240-242. Pushchino, RF: Foton-Vek.

[20] Ordchonikidze, C.G.; Ramayya, L.K.; Egorova, E.M.; Rubanovich, A.V. Toxical and genotoxical effects of silver nanoparticles on mice *in vivo*.

Abstracts of the 4-th International Conference Environmental effects of nanoparticles and nanomaterials. Vienna, 6-9th September, 2009.

[21] Ramayya, L.K.; Ordchonikidze, C.G.; Egorova, E.M.; Rubanovich, A.V. Genotoxical effects of silver nanoparticles at their action on mammalians *in vivo*. *Acta Naturae*, 2009, N3, p.109-112.

[22] Durham, A. C. H.; Ridgway, E. B. Control of Chemotaxis in *Physarum polycephalum*. *J. Cell Biol.*, 1976, v.69, p.218-223.

[23] Knowles, D. J. C.; Carlile, M. J. The Chemotactic Response of Plasmodium of the Myxomycete *Physarum polycephalum* to Sugars and Related Compounds. *J. Gen. Microbiol.*, 1978, v.108, p.17-25.

[24] Kincaid, R. L.; Mansour, T. E. Measurement of Chemotaxis in the Slime Mold Physarum polycephalum. *Exptl. Cell Res.*, 1978, v.116, p.365-375.

[25] Ueda, T.; Kobatake, Y. Chemotaxis in Plasmodia of *Physarum polycephalum*. In H. C. Aldrich & J. W. Daniel (Eds.), Cell Biology of Physarum and Didymium: Organisms, Nucleus, and Cell Cycle, 1982, v.1, p.111–143. New York: Academic Press.

[26] Beylina, S. I.; Matveeva, N. B.; Teplov, V. A. Autonomous Motive Activity and Chemotactic Behaviour of *Physarum polycephalum* Plasmodium. *Biophysics*, 1996, v.41, p.137-143.

[27] Egorova, E.M.; Revina, A.A.; Kondratieva, V.S. The mode of preparation of nanosized metallic particles. Patent RF N 2147487, priority from 01.07.1999.

[28] Egorova, E.M.; Revina, A.A. Synthesis of metallic nanoparticles in reverse micelles in the presence of quercetin. *Colloids Surfaces A*, 2000, v.168, p.87-96.

[29] Egorova, E.M.; Revina, A.A. Optical properties and sizes of silver nanoparticles in micellar solutions. *Colloid Journal* (Russian), 2002, v.64, p.334-345.

[30] Egorova, E.M.; Revina, A.A.; Rumyantzev, B.V. et al. Stable silver nanoparticles in water dispersions obtained from micellar solutions. *J. Applied Chem.* (Russian), 2002, v.75, p.1620-1625.

[31] Kamiya, N.; Kuroda, K. Studies on the Velocity Distribution of the Protoplasmic Streaming in the Myxomycete Plasmodium. *Protoplasma,* 1958, v.49, p.1-4.

[32] Wolf, K.V.; Stockem, W.; Wohlfarth-Bottermann, K.E. Cytoplasmic actomyosin fibrils after preservation with high pressure freezing. *Cell Tissue Res.*, 1981, v.217, p.479-495.

[33] Beylina, S. I.; Cieslawska, M.; Hrebenda, B.; Baranowski, Z. The Relationship between the Respiratory Rate and the Period of the

Contraction-Relaxation Cycle in Plasmodia of *Physarum polycephalum. Acta Protozoologica*, 1989, v.28, p.165-174.

[34] Wohlfarth-Botterman, K. E.; Stockem, W. Die Regeneration des Plasmalemms von Physarum polycephalum. Wilhelm Roux's Arch. *Entwicklungsmech. Org*., 1970, v.164, p.321-340.

[35] Daniel, J. W.; Baldwin, H. H. Methods of Culture for Plasmodial Myxomycetes. In J. Prescott (Ed.), Methods of Cell Physiology, 1964, v.1, p.9–41. New York: Academic Press.

[36] Camp W. G. A method of cultivating myxomycete plasmodia. *Bull. Torrey Bot. Club,* 1936, v.63, p.205-210.

[37] Teplov, V. A. Autooscillations in *Physarum* plasmodium I. Correlation between force generation and viscoelasticity during rhythmical contractions of protoplasmic strand. In M. Tazawa (Ed.), Cell Dynamics: Cytoplasmic Streaming Cell, Movement-Contraction and Migration, Cell and Organelle Division, Phototaxis of Cell and Cell Organelle, 1988, v.1, p.81-88. Wien-New York: Springer-Verlag.

[38] Kochegarov, A. A.; Beylina, S. I.; Matveeva, N. B.; Leontieva, G. A.; Zinchenko, V. P. Ionomycin and 2,5'-di(tertbutyl)-1,4,-benzohydroquinone elicit Ca^{2+}-induced Ca^{2+} release from intracellular pools in *Physarum polycephalum. Comp. Bioch. Physiol. Part A,* 2001, v.128, p.279-288.

[39] Rakoczy, L. The Myxomycete *Physarum nudum* as a Model Organism for Photobiological Studies. *Ber. Deutsch. Bot. Ges.*, 1973, v.86, p.141-164.

[40] Mohberg, J. Preparation of Spherules. In: H. C. Aldrich & J. W. Daniel (Eds.), Cell Biology of Physarum and Didymium: Differentiation, Metabolism, and Methodology, 1982, v.2, p.241-243. New York: Academic Press.

[41] Arora, S.; Jain, J.; Rajwade, J.M.; Paknikar, K.M. Cellular responses induced by silver nanoparticles: *In vitro* studies. *Toxicology letters*, 2008, v.179, p.93-100.

[42] Matveeva, N. B.; Beilina, S. I.; Teplov V. A. The Role of Phosphoinositide-3-Kinase in the Control of Shape and Directional Movement of the *Physarum polycephalum* Plasmodium. *Biophysics*, 2008, v.53, p.533-538.

[43] Chung, C. Y.; Firtel, R. A. *Dictyostelium*: A Model Experimental System for Elucidating the Pathways and Mechanisms Controlling Chemotaxis. In Conn, P. M. & Means, A. R. (Eds.), Principles of

Molecular Regulation: Signaling Mechanisms Initiated by Cell Surface Receptors part I, 2007, p.99 –114. Totowa, NJ: Humana Press Inc.

[44] Willard, S. S.; Devreotes, P. N. Signaling Pathways mediating Chemotaxis in the Social Amoeba. *Dictyostelium discoideum. Eur. J. Cell Biol.*, 2006, v.85, p.897-904.

In: Chemotaxis: Types, Clinical Significance… ISBN: 978-1-61728-495-3
Editor: T. C. Williams, pp. 189-

Chapter 7

Chlamydomonas as the Unicellular Model for Chemotaxis and Cellular Differentiation

E.V. Ermilova

Abstract

Chlamydomonas has long been one of the most successful unicellular organism for genetic and biochemical studies of the photosynthesis, organelle genomes and flagellar assembly. The availability of the new molecular genetic techniques is increasing interest in *Chlamydomonas* as a model system for research in areas like swimming behavior where it previously has not been widely exploited. The swimming behavior of *Chlamydomonas reinhardtii* is influenced by several different external stimuli including chemical attractants. Chemotaxis of the green alga is altered during gametic differentiation. Gametogenesis results in the conversion of chemotactically active vegetative cells into chemotactically inactive gametes. This experimental system offers the opportunity to study cellular behavior and differentiation at the molecular level with use of a wide range of molecular genetic approaches, including gene tagging by insertional mutagenesis, quantitative PCR and RNA interference. In this chapter I discuss recent progress in the field of chemotaxis in *Chlamydomonas*. Emphasis is placed on the signal pathways by which the two environmental cues – ammonium and light control chemotaxis and gametic differentiation.

Introduction

Motile cells and unicellular organisms exhibit a surprisingly sophisticated swimming behaviour. By monitoring changes in the chemical composition of their environment, they migrate towards food cues and mating pheromones (attractants) and flee from repellents (noxious chemicals). This behaviour, named chemotaxis, is seen in a wide range of prokaryotes (bacteria and archaea) [54, 56] and eukaryotic cells including leukocytes, macrophages and amoeboid protists, ciliated protists such as *Paramecium* and *Tetrahymena*, and flagellated cells including sperm cells and algae [12, 45, 53].

The key molecular and cellular mechanisms that underlie chemotaxis have been studied mostly on heterotrophic unicellular organisms, essentially enterobacteria [25, 5], *Dictyostelium* [79, 44], and *Paramecium* [40]. However, many molecular details of receptor function, signal processing, of regulation and control during life cycle are still poorly understood. The more recent interest in chemotaxis of phototrophic cells may give new answers to hitherto unsolved questions [21, 13]. The biflagellate green alga *Chlamydomonas reinhardtii* is well suited to molecular-genetic studies of fundamental processes in photosynthetic flagellates. We are exploiting this motile organism to investigate flagellate chemotaxis.

This unicellular alga displays positive chemotaxis towards various sugars [15, 20]. In the oogamous species *C. allensworthii*, it has been shown that the female gamete secretes a D-xylosylated quinone as a pheromone, and it is the sugar component that serves as the attractant [74]. However, the significance of the chemotaxis to sugars is unclear, since *C. reinhardtii* cannot metabolize any of the sugars, and sexual pheromones are not involved in mating in this isogamous species [39]. On the other hand, acetate, a sole carbon source utilized by *C. reinhardtii* cells during heterotrophic growth, is not a chemoattractant for them. *C. reinhardtii* vegetative cells are attracted also to the nitrogen sources, ammonium [71, 19] and nitrate/nitrite [23].

The focus of this chapter will be on chemotaxis of *C. reinhardtii* (referred to as *Chlamydomonas* for the remainder of chapter) to ammonium and nitrate. Chemotaxis of *Chlamydomonas* is altered during sexual life cycle [21, 22]. Gametogenesis results in the conversion of chemotactically active vegetative cells into chemotactically inactive gametes. Gametic differentiation itself is not of primary concern in this chapter. In-depth reviews on control of gametogenesis in *Chlamydomonas* have been published elsewhere [4, 60]. We will discuss the extrinsic cues that control chemotaxis responses during gametogenesis as well as the signal pathways through which these cues may act. Chemotaxis of vegetative cells will be considered first, followed by a discussion of its loss in gametes.

Advantages of *Chlamydomonas* as a Model Organism for the Analysis of Chemotaxis

Chlamydomonas has many advantages for biochemical, genetic and molecular approaches to understanding the mechanisms of swimming behaviour. Many of the behavioural responses of *Chlamydomonas* to signals are similar to those exhibited by other flagellates. Elucidation the mechanisms by which these signals are perceived and transmitted into swimming responses will help define regulatory strategies used by all photosynthetic cells. I have focused on the technologies that have been applied to studies of chemotaxis of *Chlamydomonas*.

Chlamydomonas is easily grown and maintained in the laboratory. Because the alga completes its sexual cycle in less then two weeks, genetic analysis is rapid. Genetic techniques such as tetrad analysis and stable diploid construction are routine for analysis of many biological processes [51, 37, 39] including chemotaxis in *Chlamydomonas* [14].

Another advantage of working with *Chlamydomonas* is the ease with which insertional mutants can be isolated. This method is based on the random integration of transforming DNA into the nuclear genome of *Chlamydomonas* [75, 38]. For efficient and stable nuclear transformation a variety of techniques have emerged, including biolistic procedure, agitation with glass beads, or electroporation [10, 47, 48]. The simplest method of introducing DNA into nuclear genome is the procedure involving glass beads, which has been used for the generation of chemotaxis mutants [20, 18, 13].

The argininosuccinate lyase 7 (*ARG7*) marker as the insertional mutagen [62] is used routinely to rescue the corresponding mutant allele (*arg7*) to prototrophy [20, 9]. This approach, however, restricts the using of recipient strains to the mutants deficient in *arg7* that, in some cases, will show relatively low chemotaxis efficiencies under conditions used for screening. Several genes can be used to select transformants: *Streptoalloteichus hindustanus ble* gene confers resistance to bleomycin [73] and *Streptomyces rimosus* aminoglycoside 3′-phosphotransferase *aphVIII* gene confers resistance to paromomycin [70]. These dominant selectable markers facilitate isolation of chemotaxis mutants. Scheme of screening for chemotaxis mutants has been developed [18, 13]. Insertion of the DNA into a functional gene results in a stable null mutant, in which the disrupted gene is 'tagged' and therefore amenable to molecular cloning [75]. Wild-type genes can be transformed into these null mutants for phenotype testing. However, null mutations in some genes of interest may have a lethal phenotype. In this case gene function can be evaluated by suppression of specific gene activities using antisense or RNAi interference (RNAi) constructs.

RNA silencing, also termed post-transcriptional gene silencing (PTGS) or RNA interference (RNAi), involves double-stranded RNA (dsRNA) intermediates that may specifically affect gene expression at the transcriptional and/or posttranscriptional levels [2, 41, 55]. A number of recent reports describe research in which tools that trigger RNA silencing via antisense or inverted repeat constructs have been developed in *Chlamydomonas* [30, 69, 42, 67]. *Chlamydomonas* possesses a single phototropine gene (*Phot*) [43]. Because *Chlamydomonas* mutants defective in *Phot* are not available, the RNAi technology was applied to test whether phototropin plays a role in the light control of gametic differentiation [42] and chemotaxis during the sexual life cycle [17].

The potential of *Chlamydomonas* as a model system has been boosted by the availability of the nearly 300,000 *Chlamydomonas* expressed sequence tags (ESTs) [11] and a draft *Chlamydomonas* genome sequence (http://www.chlamy.org). The genome of *Chlamydomonas* has revealed an unexpectedly large number of the genes encoded the high-affinity transport system (HATS) for ammonium, eight Cr*Amt1* [34]. The current evidence suggests that ammonium sensing involves ammonium transporters in some way [13]. To test the role of transport systems in ammonium/methylammonium chemotaxis, the expression patterns at different stages of sexual life cycle have been characterized with use of Real-Time PCR. Finally, the sensitive reporter genes for monitoring promoter activity in the nucleus have been developed [31, 65].

Sexual Life Cycle

Chlamydomonas offers a simple life cycle (Figure 1). In vegetative phase, cells are haploid and mitotic and they are of one of two genetically fixed mating types, designated plus (mt^{+}) and minus (mt^{-}).

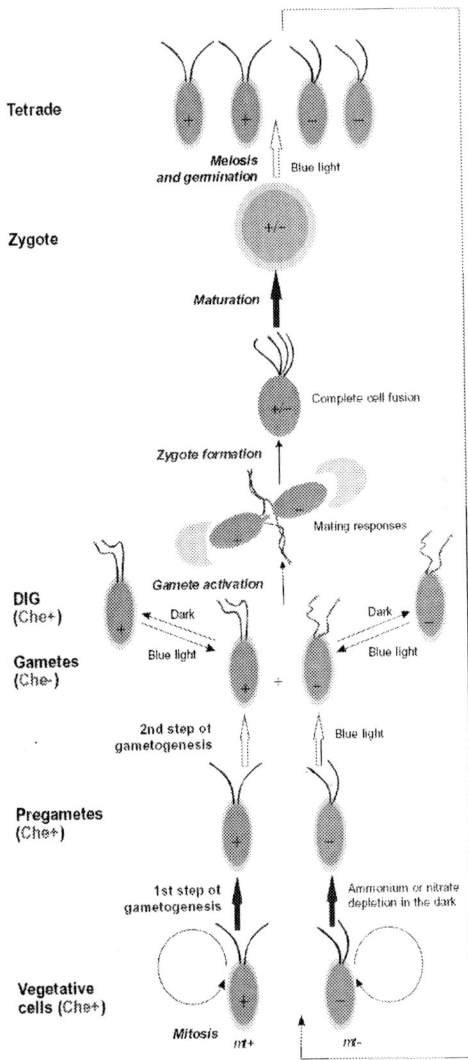

Figure 1. Life cycle of *Chlamydomonas* with indications of the mode of chemotaxis at different steps. *Che*[+] Cells exhibiting chemotaxis in response to nitrogen sources, *che*[-] cells not exhibiting chemotaxis in response to nitrogen sources, DIG dark-inactivated gametes

The sexual cycle comprises three central phases – gemetogenesis, gamete activation and cell fusion/zygote formation. The differentiation of vegetative cells into gametes initiates the sexual life cycle of *Chlamydomonas*. We use the term gametes only for those cells that have the ability to mate. *C. reinhardtii* is isogamous, which means that there are no major morphological differences between mt^+ and mt^- gametes. However, gametes are distinguished from vegetative cells biochemically, physiologically and behaviourally [63, 21].

An encounter between two gametes of opposite mating type triggers a series of events leading to fusion and the formation of a quadraflagellate zygote. Fusion itself initiates another differentiation pathway, and the meiotic zygote loses its four flagella and becomes encased in a thick zygotic cell wall. As zygotes mature, it enters a dormant state that allows the species to survive not only nitrogen starvation, but also desiccation. The cycle begins again when the appropriate environmental conditions stimulate the dormant zygote to undergo meiosis to generate four haploid progeny in unordered tetrads.

Chemotactic behaviour of the three biflagellate cell types, vegetative cells, pregametes (generated by incubation of vegetative cells without a nitrogen source in the dark) and mature gametes will be discussed. Other steps of life cycle including gamete activation and cell fusion/zygote production recently reviewed [60, 59].

Chemotaxis of Vegetative Cells

As many motile organisms, *Chlamydomonas* requires an efficient chemotactic system that allow cells to orient and migrate towards nutrients to ensure its survival. Since nitrogen is one of the macroelements that usually are limiting in many natural environments, chemotaxis to nitrogen sources appears to be an important property in vegetative cells of the alga [21, 23].

Ammonium is often a preferred source of nitrogen for many organisms, including *Chlamydomonas.* The term ammonium is used to denote both NH_3 and NH_4^+. The importance of ammonium as a nitrogen source also means that *Chlamydomonas* vegetative cells have evolved an additional adaptation such as chemotaxis that allows them to move towards ammonium [71, 19]. Chemotactic responses were tested using capillary assay [15] or a real-time, computer-based cell tracking system [20].

Chlamydomonas has been shown to possess also chemotaxis to methylammonium [13], a structural analog of ammonium routinely used for assessing ammonium transport [50]. Methylammonium, mimics the effects of ammonium by causing inactivation and/or repression of enzymes responsible for the assimilation of nitrate, nitrite and amino acids in fungi [1] and green algae [27]. *Chlamydomonas* is sensitive to methylammonium toxicity. Thus, in contrast to the chemotaxis to ammonium where vegetative cells orient themselves towards the preferred nitrogen source, in the chemotaxis to methylammonium cells do not get any benefit from attraction by this toxic compound. Motile cells may possibly sense ammonium and its analog methylammonium by the same components of the signaling pathway. So, ammonium-derivative metabolites appear not to be involved in the chemotaxis signaling pathway since methylammonium is not metabolized by *Chlamydomonas* [27].

It was shown that some *Chlamydomonas* transporters are responsible for the entry of both ammonium and methylammonium into the cells, so that ammonium inhibits competitively the methylammonium uptake activity [29]. We have found that ammonium also blocked chemotaxis to methylammonium, though methylammonium was not efficient in blocking chemotaxis to ammonium. Chemotaxis to ammonium/methylammonium might be related to the function of ammonium/methylammonium transporters and the observed inhibitions to the efficiency of the transport for each compound.

Physiological studies in *Chlamydomonas* have shown that the kinetics of ammonium uptake is biphasic and separable into a high-affinity system (HATS) and a low-affinity system (LATS) [29]. A methylammonium-resistant mutant hat1 with impaired activity of HATS for ammonium has been isolated by insertional mutagenesis [13]. The data are clearly connected to the strong inhibition of methylammonium transport in mutant strain hat1, which resulted in the complete loss of chemotaxis to ammonium/methylammonium.

Chlamydomonas exhibits active orientation with respect to a light source [81]. It is believed that photoexcitation of retinal-containing receptors results in the generation of photoreceptor current, membrane depolarization and a subsequent change in intraflagellar Ca^{2+} concentration [69]. Moreover, it has been established that the differential responses of the two flagella during phototactic turning are controlled by submicromolar concentrations of Ca^{2+} [46]. Chemotaxis is dependent upon extracellular Ca^{2+} as well [7, 19]. The chemotaxis and phototaxis signalling pathways converge, possibly at the level of transmembrane Ca^{2+} fluxes and triggering of the flagella current [16, 20]. Using ^{45}Ca we have determined that ammonium influenced the increase in cytoplasmic Ca^{2+} upon photostimulation [23]. In *Paramecium* and *Tetrahymena* attractants are shown to hyperpolarize the cell membrane, most probably by stimulating a Ca^{2+}-pump current [80]. It is probable that attractant - ammonium induces Ca^{2+} efflux and hyperpolarization of membrane in *Chlamydomonas* vegetative cells. As a result, the swimming behaviour at any point in time is due to sensory integration of photo- and chemo-signals in the form of the cell membrane potential.

A working model of how the ammonium signal is perceived and transduced during chemotactic response to ammonium is proposed (Figure 2). We suggest that the activity of some ammonium transporters is involved in the chemotactic response to ammonium, possibly by initiating of a cascade of electrical phenomena on the cell membrane and thus transducing the chemo-signal to the flagella that lead to transient changes in the flagellar beat pattern to produce chemotactic turning.

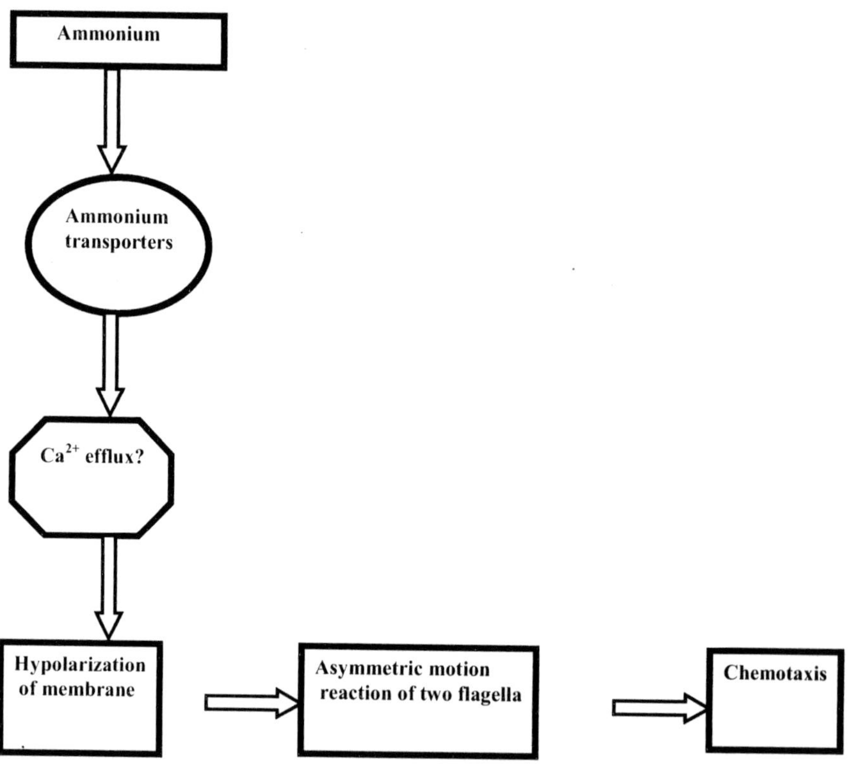

Figure 2. Proposed pathway for signal perception and transduction during chemotaxis in response to ammonium in *Chlamydomonas*

In addition to assimilating ammonium, *Chlamydomonas* can utilize nitrogen as nitrate, nitrite, urea, purines and amino acids [36]. From nitrogen sources tested only nitrate and nitrite act as attractants for *Chlamydomonas* [23]. The assimilation of nitrate involves just a few steps: uptake, reduction to nitrite, reduction to ammonium then incorporation into amino acids [26]. In *Chlamydomonas*, nitrate assimilation genes are subject to repression by ammonium [32]. Ammonum insensitive mutants for nitrate assimilation were characterized [35]. Some of them showed a deregulated pattern of *Amt1* genes expression. Nevertheless, they were found to show normal chemotaxis to ammonium and to nitrate. These events appear to be specific and independent of those involved in nitrate assimilation. The *Chlamydomonas* genome contains three different gene families (*Ntr1, Ntr2* and *Nar1*) encoding 13 putative nitrate/nitrite transporters [26]. The possible role of these transporters in the chemotacic responses remains to be understood.

Chemotaxis of Cells during Gametogenesis

The ability of unicellular organisms differentiate in response to nutrient availability is essential to their survival in a changing environment [52]. *Chlamydomonas* represents one of the model organisms to study mechanisms of sexual differentiation [4, 59]. The initial step in the sexual life cycle of *Chlamydomonas* is gametogenesis. During gametogenesis vegetative cells are transformed into mating-competent gametes. The gametic differentiation has been shown to be associated with changes not only in the cells' biochemistry and subcellular morphology [4] but also in chemotactic behavior [7, 21, 13, 23]. Mature gametes do not exhibit chemotaxis to ammonium, methylammonium, nitrate/nitrite. Switch-off of chemotaxis to these attractants in mating-competent cells is controlled by gamete-specific genes that are common for both mating-type gametes [21, 13, 23]. Loss of chemotaxis towards nitrogen sources, which are known to play a key role in the repression of the gametic differentiation, may be viewed as one of the cellular adaptations to changing environmental conditions.

Role of Ammonium in Change of Chemotaxis Mode

There is evidence that ammonium itself plays a crucial role in the repression of gametic differentiation [57]. To verify that changes in the chemotaxis mode occur mainly because of deficiencies in the ability to use ammonium, we tested the relationships between the utilization of various L-amino acids for growth and chemotaxis.

Chlamydomonas may use amino acids as sole nitrogen sources [78]. In light some amino acids may influence the switch-off chemotaxis [13]. Loss of chemotactic activity was found in wild type cells in acetate-containing nitrogen-free medium supplemented with threonine, proline and glutamate that cannot be utilized by *Chlamydomonas* cells [61]. However, when the amino acids used were alanine, serine and phenylalanine no change of chemotaxis was observed. As reported previously [61, 58], deamination of amino acids occurs extracellularly in a reaction mediated by a nonspecific L-amino-acid oxidase and requires the presence of acetate in the medium, generating ammonium and the corresponding 2-oxoacids that are not assimilated by the cells and accumulate in the medium. Arginine prevented the loss of chemotaxis even in the absence of acetate [22]. It was shown that arginine is the only amino acid, which can be transported into the cells by a specific transport system induced by nitrogen starvation [49]. Therefore, arginine can be assimilated by cells to yield intracellular ammonium. One possible reason for the results obtained was that not extracellular but intracellular ammonium completely blocked the loss of chemotaxis in the differentiating cells.

Urea also interfered with change in chemotaxis mode [22]. Urea catabolized to yield ultimately ammonium [36]. We assume that change in chemotaxis mode is triggered by a decrease in concentration of intracellular ammonium. According to our model, an intracellular threshold level of ammonium may be increased by the assimilation of alternative nitrogen sources, e.g. amino acids and urea, and this increase in concentration represses switch-off chemotaxis. These results indicate that the availability of ammonium provided by amino acids or urea but not the nitrogen-containing compounds *per se,* interfere with changes in the chemotaxis mode, and that ammonium transport activity is required to maintain an active chemotaxis. Nitrate/nitrite also block both gametic differentiation and loss of chemotaxis. Whether the regulatory networks by which ammonium and nitrate control the components responsible for gametogenesis and for switch-off chemotaxis remains to be determined.

Ammonium Transport in Gametes

Physiological studies in *Chlamydomonas* have shown that the kinetics of ammonium uptake is biphasic and separable into a high-affinity system

(HATS) and a low-affinity system (LATS) [29]. We pose the question: does switch-off chemotaxis to methylammonium/ammonium correlate with block of ability to transport them by gametes? *Chlamydomonas* is unable to metabolize methylammonium [27], avoiding interference between metabolism and transport in uptake experiments. Kinetic parameters of [^{14}C]-methylammonium uptake were determined by using the initial rate method of Lineweaver-Burk [28]. Two components were observed for methylammonium transport in gametes, LATS, with high V_{max} and K_m values (80 μmol/mg chl·h and 500 μM respectively), and HATS, with low V_{max} and K_m values (2 μmol/mg chl·h and 1.8 μM respectively). Thus, gametes express two systems for the uptake of methylammonium/ammonium.

In *Chlamydomonas* a total of eight members of the *Amt1* gene family (Cr*Amt1*.(1-8) genes) with a close homology to *AMT1* from plants has been identified as encoding the high-affinity ammonium transport system [34, 26]. To gain information on the role of the ammonium transporters in switch-off chemotaxis, comparative analysis of expression patterns for the eight *Amt1* genes in vegetative cells and gametes of wild type was performed by Real Time PCR. *Amt1*.1, *Amt1*.2, *Amt1*.4, and *Amt1*.5 with similar expression levels in ammonium-grown vegetative cells enhanced them at different extents in gametes; *Amt1*.7 is highly expressed in vegetative cells and in gametes; *Amt1*.3 has a low expression; and *Amt1*.6 shows higher expression in vegetative cells than in gametes [13]. The hypothesis can be explored that chemotaxis in gametes is blocked in transduction pathway downstream from the transport/signal perception step. However, the precise physiological role of each of the CrAMT1 transporters in chemotaxis will become clearer once we have more information about their subcellular localizations.

Role of Light in Change of Chemotaxis Mode

The differentiation of vegetative cells into sexually mature cells in *Chlamydomonas* may proceed in two steps (Figure 1) [4]. The depletion of an utilizable nitrogen source induces the first step that results in sexually noncompetent gametes [66, 3]. In the second step, these cells may be converted into mating-competent gametes by irradiation with light [76]. The light-competent intermediates in gamete formation were termed pre-gametes. Since three types of cells during gametogenesis are well defined by physiological and molecular criteria [77, 76], a solid basis for a comparative behavioral analysis of each cell type is available.

We have shown that change of chemotactic behavior may be divided into the light-independent and light-dependent steps, like the formation of competence ability [21, 17]. Thus, vegetative cells and mating incompetent pre-gametes exhibit chemotaxis towards ammonium. Irradiation of pre-gametes results in a loss of chemotaxis and the gaining of mating competence (Figure 1). Incubation of these gametes in the dark resulted in their regaining of chemotactic activity; reillumination again in its loss [17].

When light of different wavelengths was employed, blue light was shown to be most effective in changing chemotactic behavior [17]. Two classes of blue-light receptors have been identified in recent years in higher plants, the cryptochromes and the phototropins [6, 8]. Single genes for both blue-light receptors have also been identified in *Chlamydomonas* [43, 72]. The cryptochrome of *Chlamydomonas* appears to be light-labile because the protein disappears after irradiation of the cells [64]. Its function has not yet been elucidated. Phototropin in this alga has a structure typical for that of other members of the phototropin family [43]. The blue-light receptor was shown to control progression of the sexual life cycle of this green alga [42].

Using strains with reduced levels of the blue-light receptor phototropin (RNAi20 and RNAi30), a correlation between intracellular levels of this photoreceptor and the shut-off of chemotaxis became evident. As shown in Figure 3a, both strains exhibited reduced levels of phototropin, although the PHOT levels of the RNAi20 strain were distinctly lower than those of RNAi30. Thus, the kinetics of the loss of chemotaxis to ammonium approximately correlates with the level of PHOT protein detected by specific antibodies. While a reduction in phototropin molecules lead to a delay in switch-off of chemotaxis, this defect could partially be compensated by the application of higher fluence rates (Figure 3b). Reduced phototropin levels not only affected the loss of chemotactic behavior upon illumination of pre-gametes but also that of gametes that had been shifted into the dark, resulting in a reappearance of chemotaxis [17]. Lower levels of phototropin in these cells also delayed the inactivation of chemotaxis. In summary, these results suggest that phototropin is the photoreceptor by which blue light inactivates chemotaxis towards ammonium in nitrogen-starved cells (Figure 4).

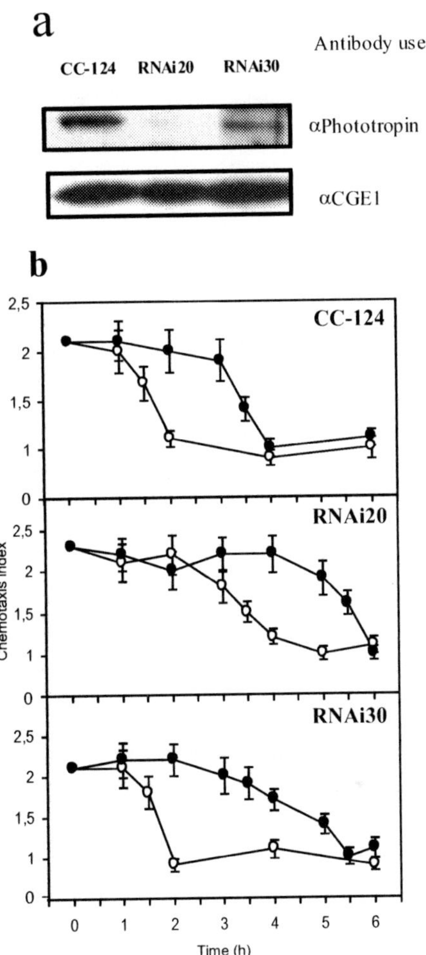

Figure 3. a, b. Effect of a reduction in phototropin levels on the light-induced loss of chemotaxis of *Chlamydomonas* towards ammonium.
a. Relative amounts of phototropin in different transformants harboring RNAi constructs as compared to wild-type strain CC-124. For this assay, an antibody directed against *Chlamydomonas* phototropin (Huang et al., 2002) was used. For a loading control, an antibody that reacts with the co-chaperone CGE1 (Schroda et al., 2001) was employed. b. Chemotaxis of strains CC-124, RNAi20 and RNAi30. Pre-gametes, generated from synchronously growing cells by incubation in nitrogen-free medium in the dark for 18 h, were shifted at time 0 into white light of fluence rates 60 $\mu molm^{-2}s^{-1}$ (With kind permission from Springer Science+Business Media: <Planta, Phototropin plays a crucial role in controlling changes in chemotaxis during the initial phase of the sexual life cycle in *Chlamydomonas*, 219, 2004, 420-427, Elena V. Ermilova, Zhanna M. Zalutskaya, Kaiyao Huang, Christoph F. Beck, figure 4>).

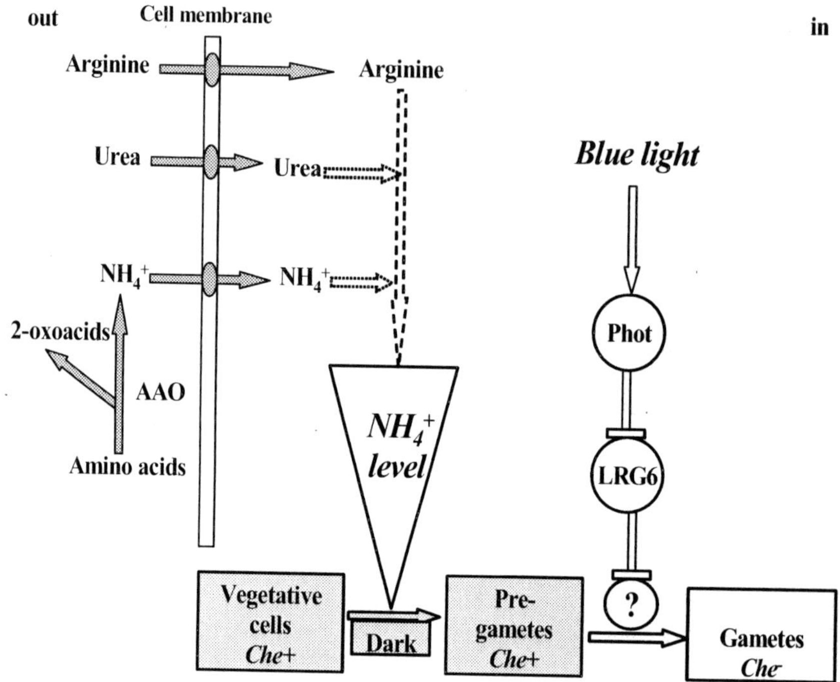

Figure 4. A model to explain the intracellular interactions of signaling pathways by which ammonium and blue light control chemotaxis mode towards nitrogen sources in *Chlamydomonas* during vegetative cells-to-gamete conversion. *Che*$^+$ Cells exhibiting chemotaxis in response to nitrogen sources, *che*$^-$ cells not exhibiting chemotaxis in response to nitrogen sources, AAO L-amino acids oxidase, PHOT phototropin.

Chemotaxis to methylammonium and to nitrate/nitrite was also a property of vegetative cells and pre-gametes but not of gametic cells [13, 23]. The RNAi20 strain showed an attenuated inactivation of chemotaxis to methylammonium and to nitrate/nitrite. Thus, phototropin appears to be the photoreceptor that perceives the light, which induces loss of chemotactic responses to methylammonium and to nitrate/nitrite during the sexual life cycle.

Three *lrg* mutants that exhibit light-independent sexual differentiation [33] exhibited loss of chemotaxis to methylammonium/ammonium and nitrate/nitrite even in the dark [21, 24, 23]. Light independence of pregamete-to-gamete conversion was also observed in mutant *lrg6*. The *LRG6* gene product was suggested to be a negative regulator in the signaling pathway that controls gamete formation [9]. This mutant showed loss of chemotaxis to these attractants in the absence of light. The sharing of photoreceptor and signal transduction components as well as similar temporal patterns observed for changes in chemotaxis towards ammonium and nitrate/nitrite suggest an integration of the signalling pathways that control these responses, supporting the hypothesis that chemotaxis in gametes is blocked in transduction pathway downstream from the signal perception step. A model that explains how ammonium and blue light may control the chemotaxis during the convertion of vegetative cells into gametes is presented in Figure 4.

The data also suggest a tight coupling between changes in chemotaxis towards nitrogen sources and the formation of gametes [21, 24]. Why does *Chlamydomonas* exhibit this intriguing coupling of light control for gamete formation and the disappearance of chemotaxis towards the nitrogen sources? We surmise that gametes are no longer in search for a nitrogen source. Since gametes in nature will have to mate at locations accessible to light – only this way the non-motile zygotes formed will have a chance to germinate in a process that is strictly dependent on light [33] – a search for a nitrogen source may divert them to places not accessed by light. During the night, gametes reverting to dark-inactivated gametes, may move towards an ammonium or nitrate/nitrite source – but phototaxis towards light can reverse this movement at the beginning of the day. Seen from this point of view, the coordinated but opposite regulation of chemotaxis towards ammonium or nitrate/nitrite and gamete formation may be of advantage for the algae.

Conclusion

Recent discoveries of behaviour in *Chlamydomonas* have revealed the importance of chemotaxis not only in migration towards food cues, but also in development processes. Inorganic nitrogen is available to *Chlamydomonas* vegetative cells and pre-gametes (non-competent gametes) as nutrient and attractant in both cationic and anionic forms, ammonium and nitrate respectively. Mature gametes do not exhibit chemotaxis towards these nitrogen sources, which are known to repress the gametic differentiation. The two environmental cues that control this differentiation – the scarcity of a nitrogen source and blue light – are signals that play important roles in the switch-off of chemotaxis as well. We are only beginning to understand the mechanisms of chemotaxis at molecular level, and we are left with major questions such as: (i) how ammonium and nitrate are sensed by vegetative cells and pre-gametes (ii) what is the role of known and newly identified ammonium and nitrate transporters in nitrogen sensing? (iii) which are the regulatory pathways that link ammonium and nitrate sensing and flagellar responses during chemotaxis? (iv) which signal transduction elements are activated through blue-light perception by phototropin? Future application of the developing molecular technologies to the analysis of chemotaxis will be essential for solving such questions. In addition to understanding basic processes involved in the chemosensory signal reception and transduction in a simple plant organism, the solutions may grant new insights into fundamental principles of nitrogen sensing in photosynthetic eukaryotes.

Acknowledgments

I would like to thank the Russian Foundation for Basic Research (Grant 07- 04-00277) and INTAS (Grant 05-1000008-8004) for continued support of our work on chemotaxis in *Chllamydomonas.*

References

[1] Arst, HN; Cove, DJ. Methylammonium resistance in *Aspergillus nidulans*. *J Bacteriol*, 1969, 98, 1284-1293.

[2] Baulcombe, D. RNA silencing in plants. *Nature*, 2004, 431, 356-363.

[3] Beck, CF; Acker, A. Gametic differentiation of *Chlamydomonas reinhardtii*. *Plant Physiol*, 1992, 98, 822-826.

[4] Beck, CF; Haring, M. Gametic differentiation of *Chlamydomonas*. *Int Rev Cytol*, 1996, 168, 259-302.

[5] Bourret, RB; Stock, AM. Molecular information processing: lessons from bacterial chemotaxis. *J. Biol. Chem*, 2002, 277, 9625-9628.

[6] Briggs, WR; Beck, CF; Cashmore, AR et al. The phototropin family of photoreceptors. *Plant Cell*, 2001, 6, 993-997.

[7] Byrne, TE; Wells, MR; Johnson, CH. Circadian rhythms of chemotaxis and of methylammonium uptake in *Chlamydomonas reinhardtii*. *Plant Physiol*, 1992, 98, 879-886.

[8] Christie, JM; Briggs, WR. Blue light sensing in higher plants. *J Biol Chem*, 2001, 276, 11457-11460.

[9] Dame, G; Gloecker, G; Beck, CF. Knock-out of a putative transporter results in altered blue-light signalling in *Chlamydomonas. Plant J*, 2002, 31, 577-587.

[10] Debuchy, R; Purton, S; Rochaix, J-D. The argininosuccinate lyase gene of *Chlamydomonas reinhardtii*: an important tool for nuclear transformation and for cor-relating the genetics and molecular maps of the *ARG7* locus. *EMBO J*, 1989, 8, 8803-8807.

[11] Eberhard, S; Jain, M; Im, CS; Pollock, S; Shrager, J; Lin, Y; Peek, AS; Grossman, AR. Generation of an oligonucleotide array for analysis of gene expression in *Chlamydomonas reinhardtii. Curr Genet*, 2006, 49, 106-124.

[12] Eisenbach, M. Sperm chemotaxis. *Rev Reprod*, 1999, 4, 56-66

[13] Ermilova, E; Nikitin, M; Fernandez, E. Chemotaxis to ammonium/ methylammonium in *Chlamydomonas reinhardtii*: the role of transport systems for ammonium/methylammonium. *Planta*, 2007, 226, 1323-1332.

[14] Ermilova, EV; Chekunova, EM; Zalutskaya, ZhM; Krupnov, KR; Gromov, BV. Isolation and characterization of chemotactic mutants of *Chlamydomonas reinhardtii. Curr Microbiol*, 1996, 32, 357-359

[15] Ermilova, EV; Zalutskaya, ZhM; Gromov, BV. Chemotaxis towards sugars in *Chlamydomonas reinhardtii. Curr Microbiol*, 1993, 27, 47-50.

[16] Ermilova, EV; Zalutskaya, ZhM; Gromov, BV. Chemotaxis and its correlation with photoresponse in *Chlamydomonas reinhardtii* strain with negative phototaxis. *Russian Biology Bulletin*, 1997, 4, 500-503.

[17] Ermilova, EV; Zalutskaya, ZhM; Huang, K; Beck, CF. Ptototropin plays a crucial role in controlling changes in chemotaxis during the initial phase of the sexual life cycle in *Chlamydomonas. Planta*, 2004a, 219, 420-427.

[18] Ermilova, EV; Nikitin, MM; Lapina, TV; Zalutskaya, ZM. *Cha1*, a DNA insertional transformant of the green alga *Chlamydomonas reinhardtii* with altered chemotaxis to ammonium. *Protistology*, 2004b, 3, 230-240.

[19] Ermilova, EV; Zalutskaya, ZhM; Munnik, T; van den Ende; H, Gromov, BV. Calcium in the control of chemotaxis in *Chlamydomonas. Biologia, Bratislava*, 1998, 53, 577-581.

[20] Ermilova, EV; Zalutskaya, ZhM; Gromov, BV; Häder, D-P; Purton, S. Isolation and characterisation of chemotactic mutants of *Chlamydomonas reinhardtii* obtained by insertional mutagenesis. *Protist*, 2000, 151, 127-137.

[21] Ermilova, EV; Zalutskaya, ZhM; Lapina, T; Nikitin, MM. Chemotactic behavior of *Chlamydomonas reinhardtii* is altered during gametogenesis. *Current Microbiol*, 2003a, 46, 261-264.

[22] Ermilova, EV; Zalutskaya, ZhM; Lapina, T; Nikitin, MM. Effects of nitrogen-containing compounds on change in chemotaxis mode during gametogenesis of *Chlamydomonas reinhardtii. Protistology*, 2003b, 3, 9-14.

[23] Ermilova, EV; Nikitin, MM; Zalutskaya, ZhM; Lapina, TV; Fernandez, E. Chemotaxis to ammonium and nitrate in *Chlamydomonas reinhardtii*: molecular components of reception and transport systems. In: Podlubnaya ZA, Malyshev SL. *Biological motility: achievements and perspectives*. Pushino: Foton-Vek; 2008, 136.

[24] Ermilova, EV; Zalutskaya, ZhM, Baibus, DM; Beck, CF. Characterization of phototropin-controlled signaling components that regulate chemotaxis towards ammonium in *Chlamydomonas. Protistology*, 2006, 4, 301-310.

[25] Falke, JJ; Bass, RB; Butler, SL; Chervitz, SA; Daniellson, MA. The two-component signaling pathway of bacterial chemotaxis: a molecular view of signal transduction by receptors, kinases, and adaptation enzymes. *Annu Rev Cell Dev Biol*, 1997, 13, 457-512.

[26] Fernandez, E; Galvan, A. Inorganic nitrogen assimilation in *Chlamydomonas*. *J Exp Bot*, 2007, 58, 2279-2287.

[27] Franco, AR; Cárdenas, J; Fernández, E. Ammonium (methylammonium) is the co-repressor of nitrate reductase in *Chlamydomonas reinhardtii. FEBS Lett*, 1984, 176, 453-456.

[28] Franco, AR; Cárdenas, J; Fernández, E. A mutant of *Chlamydomonas reinhardtii* altered in the transport of ammonium and methylammonium. *Mol Gen Genet*, 1987, 206, 414-418.

[29] Franco, AR; Cárdenas, J; Fernández, E. Two different carriers transport both ammonium and methylammonium in *Chlamydomonas reinhardtii. J Biol Chem*, 1988, 263, 14039-14043.

[30] Fuhrmann, M; Stahlberg, A; Govorunova, E; Rank, S; Hegemann, P. The abundant retinal protein of the *Chlamydomonas* eye is not the photoreceptor for phototaxis and photophobic responses. *J Cell Sci*, 2001, 114, 3857-3863.

[31] Fuhrmann, M; Hausherr, A; Ferbitz, L; Schodl, T; Heitzer,, M; Hegemann, P. Monitoring dynamic expression of nuclear genes in *Chlamydomonas reinhardtii* by using a synthetic luciferase reporter gene. *Plant Mol Biol*, 2004, 55, 869-881.

[32] Galvan, A; Fernandez, E. Eukaryotic nitrate and nitrite transporters. *Cell Mol Life Sci*, 2001, 58, 225-233.

[33] Gloeckner, C; Beck, CF. Genes involved in light control of sexual differentiation in *Chlamydomonas reinhardtii. Genetics*, 1995, 141, 937-943.

[34] González-Ballester, D; Camargo, A; Fernández, E. Ammonium transporter genes in *Chlamydomonas*: the nitrate-specific regulatory gene *Nit2* is involved in *Amt1;*1 expression. *Plant Mol Biol*, 2004, 56, 863-878.

[35] Gonzales-Ballester, D; de Montaigu, A; Higuerra, JJ; Galvan, A; Fernandez, E. Functional genomics of the regulation of the nitrate assimilation pathway in *Chlamydomonas. Plant Physiol*, 2005, 137, 522-533.

[36] Grossman, A; Takahashi, H. Macronutrient utilization by photosynthetic eukaryotes and the fabric of interactions. *Annu Rev Plant Physiol Plant Mol. Biol*, 2001, 52, 163-210.

[37] Grossman, A. *Chlamydomonas reinhardtii* and photosynthesis: genetics to genomics. *Curr Opin Plant Biol*, 2000, 3, 132-137.

[38] Gumpel, NJ; Purton, S. Playing tag with *Chlamydomonas. Trends Cell Biol*, 1994, 4, 299-301.

[39] Harris, EH. *Chlamydomonas* as a model organism. *Annu Rev Plant Physiol Plant Mol Biol*, 2001, 53, 363-406.

[40] Hennessey, TM; Kim, MY; Satir, BH. Lysozyme acts as a chemorepellent and secretagogue in *Paramecium* by activating a novel receptor-operated Ca^{2+} conductance. *J Membr Biol*, 1995, 148, 13-25.

[41] Herr, AJ. Silence is green. *Biochem Soc Trans,* 2004, 32, 946-951.

[42] Huang, K; Beck, CF. Phototropin is the blue-light receptor that controls multiple steps in the sexual life cycle of the green alga *Chlamydomonas reinhardtii. Proc Natl Acad Sci USA*, 2003, 100, 6269-6274.

[43] Huang, K; Merkle, T; Beck, CF. Isolation and characterization of a *Chlamydomonas* gene that encodes a putative blue-light photoreceptor of the phototropin family. *Physiol Plant*, 2002, 115, 613-622.

[44] Insall, I; Andrew, N. Chemotaxis in *Dyctiostelium*: how to work straight using parallel pathways. *Curr Opin Microbiol*, 2007, 10, 578-581.

[45] Jaenicke, L. One hundred and one years of chemotaxis - Pfeffer, pheromones, and fertilization. *Bot Acta*, 1988, 101, 149-159.

[46] Kamiya, R; Witman, GB. Submicromolar levels of calcium control the balance of beating between the two flagella in the demembraned models of *Chlamydomonas. J Cell Biol*, 1984, 98, 97-107.

[47] Kindle, KL. High-frequency nuclear transformation of *Chlamydomonas reinhardtii*. *Proc Natl Acad Sci USA*, 1990, 87, 1228-1232.

[48] Kindle, KL; Schnell, RA; Fernandez, E; Lefebvre, PA. Stable nuclear transformation of *Chlamydomonas* using the *Chlamydomonas* gene for nitrate reductase. *J Cell Biol*, 1989, 109, 2589-2601.

[49] Kirk, DL; Kirk, MM. Carrier-mediated uptake of arginine and urea by *Chlamydomonas reinhardtii*. *Plant Physiol*, 1978, 61, 556-560.

[50] Kleiner, D. Bacterial ammonium transport. *FEMS Microbiol Rev*, 1985, 32, 87-100.

[51] Lefebvre, PA; Silflow, CD. *Chlamydomonas*: The cell and its genomes. *Genetics*, 1999, 151, 9-14.

[52] Lengeler, KB; Davidson, RC; D'souza, C; Toshiaki, H; Shen, W-C; Wang, P; Pan, X; Waugh, M; Heitman, J. Signal transduction cascades regulating fungal development and virulence. *Microbiol Mol Biol Rev*, 2000, 64, 746-785.

[53] Maier, I. Gamete orientation and induction of gametogenesis by pheromones in algae and plants. *Plant Cell Environ*, 1993, 16, 891-907.

[54] Manson, MD; Armitage, JP; Hoch, JA; Macnab, RM. Bacterial locomotion and signal transduction. *J Bacteriol*, 1998, 180, 1009-1022.

[55] Martienssen, RA; Zaratiegui, M; Goto, DB. RNA interference and heterochromatin in the fission yeast *Schizosaccharomyces pombe*. *Trends Genet*, 2005, 21, 450-456.

[56] Marwan, W; Oesterhelt, D. Archaeal vision and bacterial smelling. *ASM News*, 2000, 66, 83-89.

[57] Matsuda, Y; Shimada, T; Sakamoto, Y. Ammonium ions control gametic differentiation and dedifferentiation in *Chlamydomonas reinhardtii*. *Plant Cell Physiol*, 1992, 33, 909-914.

[58] Muños-Blanco, J; Hidalgo-Martinez, J; Cárdenas, J. Extracellular deamination of L-amino acids by *Chlamydomonas reinhardtii* cells. *Planta*, 1990, 182, 194-198.

[59] Pan, J; Misamore, MJ; Wang, Q; Snell, WJ. Protein transport and signal transduction during fertilization in *Chlamydomonas*. *Traffic*, 2003, 4, 452-459.

[60] Pan, J; Snell, WJ. Signal transduction during fertilization in the unicellular green alga, *Chlamydomonas*. *Curr Opin Microbiol*, 2000, 3, 596-602.

[61] Piedras, P; Pineda, M; Muños-Blanco, J; Cárdenas J. Purification and characterization of an L-amino acids oxidase from *Chlamydomonas reinhardtii* cells. *Planta*, 1992, 188, 13-18.

[62] Purton, S; Rochaix, J-D. Characterization of the *ARG7* gene of *Chlamydomonas reinhardtii* and its application to nuclear transformation. *Eur J Phycol*, 1995, 30, 141-148.

[63] Quarmby, LM. Signal transduction in the sexual life of *Chlamydomonas*. *Plant Mol Biol*, 1994, 26, 1271-1287.

[64] Reisdorph, NA; Small, GD. The *CPH1* gene of *Chlamydomonas reinhardtii* encodes two forms of cryptochrome whose levels are controlled by light-induced proteolysis. *Plant Physiol*, 2004, 134, 1546-1554.

[65] Ruecker, O; Zillner, RG-F; Heitzer, M. Gaussia-luciferase as a sensitive reporter gene for monitoring promoter activity in the nucleus of the green alga *Chlamydomonas reinhardtii*. *Mol Genet Genomics*, 2008, 280, 153-162.

[66] Sager, R; Granick, S. Nutritional control of sexuality in *Chlamydomonas reinhardtii*. *J. Gen. Physiol*, 1954, 37, 729-742.

[67] Schroda, M. RNA silencing in *Chlamydomonas*: mechanisms and tools. *Curr Genet*, 2006, 49, 69-84.

[68] Schroda, M; Vallon, O; Whitelegge, J; Beck, CF; Wollman, F-A. Identification and characterization of a chloroplast GrpE homolog. *Plant Cell*, 2001, 13, 2823-2839.

[69] Sineshchekov, OA; Jung, K-H; Spudich, JL. Two rhodopsins mediate phototaxis to low- and high-intensity light in *Chlamydomonas reinhardtii*. *Proc Natl Acad Sci* USA, 2002, 99, 8689-8694.

[70] Sizova, I; Fuhrmann, M; Hegemann, P. A *Streptomyces rimosus aphVIII* gene coding for a new type phosphotransferase provides stable antibiotic resistance to *Chlamydomonas reinhardtii*. *Gene*, 2001, 277, 221-229.

[71] Sjoblad, RP; Frederikse, PH. Chemotactic responses of *Chlamydomonas reinhardtii*. *Mol Cell Biol*, 1981, 1, 1057-1060.

[72] Small, GD; Min, B; Lefebvre, PA. Characterization of *Chlamydomonas reinhardtii* gene encoding a protein of the DNA photolyase/blue light photoreceptor family. *Plant Mol Biol*, 1995, 28, 443-454.

[73] Stevens, DR; Rochaix, JD; Purton, S. The bacterial phleomycin resistance gene ble as a domonant selectable marker in *Chlamydomonas*. Mol Gen Genet, 1996, 251, 23-30.

[74] Takanashi, S; Mori, K. Synthesis of the analogues of lurlenic acid with a modified sugar part: *Chlamydomonas* responds only to D-xyloside. *Liebigs Ann-Recl*, 1997, 6, 1081-1084.

[75] Tam, LW; Lefebvre, PA. Cloning of flagellar genes in *Chlamydomonas reinhardtii* by DNA insertional mutagenesis. *Genetics,* 1993, 135, 375-384.

[76] Treier, U; Fuchs, S; Weber, M; Wakarchuk, WW; Beck, C. Gametic differentiation in *Chlamydomonas reinhardtii*: light dependence and gene expression patterns. *Arch Microbiol*, 1989, 152, 572-577.

[77] Treier, U; Beck, CF. Changes in gene expression patterns during the sexual life cycle of *Chlamydomonas reinhardtii. Physio. Plant*, 1991, 83, 633-639.

[78] Vallon, O; Bulté, L; Kuras, R; Olive, J; Wollman, FA. Extensive accumulation of an extracellular L-amino acid oxydase during gametogenesis of *Chlamydomonas reinhardtii. Eur J Biochem*, 1993, 215, 351-360.

[79] Van Haastert, PJM; Devreotes, PN. Chemotaxis: signalling the way forward. *Nat Rev Mol Cell Biol*, 2004, 5, 626-634.

[80] Van Houten, JL. Chemosensory transduction in eukaryotic microorganisms: trends for neuroscience? *Trends Neurosci*, 1994, 17, 62-71.

[81] Witman, GB. *Chlamydomonas* phototaxis. *Trends Cell Biol*, 1993, 3, 403-408.

Index

A

acetate, 190, 198
acid, 198, 209, 210
activation, 194
adaptation, 194, 206
alanine, 198
algae, 190, 195, 203, 208
allele, 191
alternative, 198
amino, 195, 197, 198, 202, 208, 210
amino acid, 195, 197, 198, 202, 208, 210
amino acids, 195, 197, 198, 202, 208
ammonium, xii, 189, 190, 192, 194, 195, 196, 197, 198, 199, 200, 201, 202, 203, 205, 206, 208
amoeboid, 190
antibiotic resistance, 209
antibody, 201
application, 200, 204, 209
archaea, 190
arginine, 198, 208
assimilation, 195, 197, 198, 206, 207
availability, xi, 189, 192, 197, 198

B

bacteria, 190
bacterial, 204, 206, 208, 209
beating, 207
behavior, xi, 189, 197, 200, 206
biochemistry, 197
biological processes, 191

C

Ca^{2+}, 195
calcium, 207
capillary, 194
carbon, 190
cell, 194, 196, 200, 208
cell fusion, 194
changing environment, 197
chemical composition, 190
chemicals, 190
chemoattractant, 190
chemotaxis, xii, 189, 190, 191, 192, 193, 194, 195, 196, 197, 198, 199, 200, 201, 202, 203, 204, 205, 206, 207
chloroplast, 209
competence, 200
components, 195, 198, 199, 203, 206
composition, 190
compounds, 198, 206
concentration, 195, 198
construction, 191
control, xii, 189, 190, 192, 198, 200, 201, 202, 203, 205, 207, 208, 209

conversion, xii, 189, 190, 202
correlation, 200, 205
coupling, 203
cues, xii, 189, 190, 203

D

depolarization, 195
desiccation, 194
differentiation, xii, 189, 190, 192, 194, 197, 198, 199, 202, 203, 204, 207, 208, 210
diploid, 191
DNA, 191, 192, 205, 209, 210
draft, 192

E

electroporation, 191
encoding, 197, 199, 209
environment, 190
environmental conditions, 194, 197
enzymes, 195, 206
eukaryotes, 204, 207
eukaryotic cell, 190
expressed sequence tag, 192

F

family, 199, 200, 204, 207, 209
fertilization, 207, 208
food, 190, 203
fungal, 208
fungi, 195
fusion, 194

G

gamete, 190, 194, 197, 200, 202, 203
gametes, xii, 189, 190, 193, 194, 197, 199, 200, 201, 203
gametogenesis, 190, 197, 198, 200, 206, 208, 210
gene, xii, 189, 191, 192, 197, 199, 202, 205, 206, 207, 208, 209, 210
gene expression, 192, 205, 210
gene silencing, 192
generation, 191, 195
genes, 191, 192, 197, 199, 200, 206, 207, 210
genetics, 205, 207
genome, 191, 192, 197
genomes, xi, 189, 208
genomics, 207
glass, 191
glutamate, 198
growth, 190, 198

H

haploid, 193, 194
heterochromatin, 208
heterotrophic, 190
homolog, 209
homology, 199
hypothesis, 199, 203

I

illumination, 200
inactivation, 195, 200, 202
inactive, xii, 189, 190
incubation, 194, 201
independence, 202
induction, 208
information processing, 204
inhibition, 195
integration, 191, 196, 203
interactions, 202, 207
interference, xii, 189, 192, 199, 208
ions, 208
irradiation, 200
isolation, 192

J

Jung, 209

K

kinases, 206
kinetics, 195, 199, 200

L

lead, 196, 200
leukocytes, 190
life cycle, 190, 192, 193, 194, 197, 200, 201, 202, 205, 207, 210
locomotion, 208
locus, 205
luciferase, 206, 209

M

macrophages, 190
meiosis, 194
metabolism, 199
metabolites, 195
microorganisms, 210
migration, 203
mitotic, 193
model system, xi, 189, 192
models, 207
molecules, 200
morphological, 194
morphology, 197
mutagenesis, xii, 189, 195, 205, 210
mutant, 191, 195, 202, 206
mutants, 191, 192, 197, 202, 205
mutations, 192

N

NA, 209
natural, 194
natural environment, 194
neuroscience, 210
nitrate, 190, 195, 197, 198, 202, 203, 206, 207, 208
nitrogen, 190, 193, 194, 195, 197, 198, 199, 200, 201, 202, 203, 206
nuclear, 191, 205, 206, 208, 209
nuclear genome, 191
nucleus, 192, 209
nutrient, 197, 203
nutrients, 194

O

organelle, xi, 189
organism, xi, 189, 190, 204, 207
orientation, 195, 208

P

pathways, xii, 189, 191, 195, 203, 204, 207
PCR, xii, 189, 192, 199
perception, 196, 199, 203, 204
phenotype, 192
phenylalanine, 198
pheromone, 190
photoexcitation, 195
photoreceptor, 195, 200, 202, 206, 207, 209
photoreceptors, 204
photoresponse, 205
photosynthesis, xi, 189, 207
photosynthetic, 190, 191, 204, 207
phototaxis, 195, 203, 205, 206, 209, 210
phototrophic, 190
physiological, 199, 200
plants, 199, 200, 204, 208
play, 197, 203
pre-game, 200, 202, 203
production, 194
progeny, 194
prokaryotes, 190
promoter, 192, 209
property, 194, 202
protein, 200, 206, 209
proteolysis, 209
purines, 197

Q

quinone, 190

R

RA, 208
random, 191
range, xii, 189, 190
RB, 204, 206
reception, 204, 206

receptors, 195, 200, 206
reduction, 197, 200, 201
regulation, 190, 203, 207
relationships, 198
repression, 195, 197, 198
research, xi, 189, 192
resistance, 191, 204, 209
rhythms, 204
RNA, xii, 189, 192, 204, 208, 209
RNAi, 192, 201
Russian, 204, 205

S

scarcity, 203
search, 203
sensing, 192, 204
series, 194
serine, 198
sexuality, 209
sharing, 202
signal transduction, 203, 204, 206, 208
signaling, 195, 202, 206
signaling pathway, 195, 202, 206
signaling pathways, 202
signalling, 195, 203, 205, 210
signals, 191, 196, 203
solutions, 204
species, 190, 194
sperm, 190
starvation, 194, 198
strain, 195, 200, 201, 202, 205
strains, 191, 200, 201
strategies, 191
Streptomyces, 191, 209
sugar, 190, 209
sugars, 190, 205
suppression, 192
survival, 194, 197
systems, 192, 199, 205, 206

T

technology, 192
temporal, 203
tetrad, 191
threonine, 198
threshold, 198
threshold level, 198
time, 194, 196, 201
toxic, 195
toxicity, 195
transcriptional, 192
transduction, 196, 199, 203, 204, 208, 209, 210
transformation, 191, 205, 208, 209
transmembrane, 195
transport, 192, 195, 198, 199, 205, 206, 208
triggers, 194

U

urea, 197, 198, 208

V

values, 199
virulence, 208
vision, 208

W

wavelengths, 200
wild type, 198, 199

Y

yeast, 208
yield, 198

Z

zygote, 194
zygotes, 194, 203